CABI CONCISE

WASTEWATER TREATMENT

Wastewater generated from industrial processing is a major source of environmental pollution that is difficult and costly to manage. With many industries thus producing wastewater containing toxic compounds or harmful pathogens, it is mandatory to treat this wastewater to avoid contaminating sources of running clean water and impacting aquatic ecosystems. Different wastes, such as heavy metals, chemicals and organic matter, also present different challenges that require specific technological solutions.

This series examines the range of wastewater contaminants generated by various processing industries and how these can be effectively treated to ensure human health and environmental stability. The series thus provides solutions for wastewater treatment across different industries based on the most up-to-date technological and scientific advances to ensure environmental and economic sustainability.

Series Editor
Neha Srivastava,
Department of Chemical Engineering and Technology, Indian Institute of Technology (BHU), Varanasi, India
sri.neha10may@gmail.com

Microbe-assisted Remediation of Heavy Metals in Wastewater

Manikant Tripathi

Biotechnology Program, Dr Rammanohar Lohia Avadh University, Ayodhya-224001, Uttar Pradesh, India

Sukhminderjit Kaur

Department of Biotechnology, University Institute of Biotechnology, Chandigarh University, Mohali, Punjab 140413, India

CABI

CABI is a trading name of CAB International

CABI	CABI
Nosworthy Way	200 Portland Street
Wallingford	Boston
Oxfordshire OX10 8DE	MA 02114
UK	USA

Tel: +44 (0)1491 832111

T: +1 (617)682-9015

E-mail: info@cabi.org

E-mail: cabi-nao@cabi.org

Website: www.cabi.org

A catalogue record for this book is available from the British Library, London, UK.

ISBN-13: 9781836993940 (hardback)
9781836993957 (paperback)
9781836993964 (ePDF)
9781836993971 (ePub)

DOI: 10.1079/9781836993971.0000

Commissioning Editor: Jamie Lee
Editorial Assistant: Theresa Regueira
Production Editor: Theresa Regueira

Typeset by Exeter Premedia Services Pvt Ltd, Chennai, India
Printed in the USA

Contents

Preface

Heavy metal pollution is one of the major environmental issues, and arrives in the ecosystem from various sources such as wastes generated from different industries, domestic activities and natural processes. These pollutants are toxic at high concentration and pose negative impacts on biodiversity including to humans, animals and plants. It is necessary to detoxify or remediate heavy metal from contaminated environments for public health and environmental sustainability. Conventional technologies are not cost effective, and can cause secondary pollution. Biotechnological strategies using microbes like bacteria, fungi and algae for remediation of heavy metals are cost effective and environmentally friendly solutions.

This book explores the potential of microorganisms along with advanced strategies in mitigating heavy metal pollution for a sustainable future. The five chapters begin with basic concepts of heavy metals and their ecotoxicological impacts, moving on to microbial bioremediation of heavy metals with detailed discussions of bacterial, fungal and algae-based remediations and their mechanisms. This book covers the latest research on the impacts of toxic heavy metals on ecosystems, up-to-date understanding of microbial applications, along with technological advancements like nanotechnology and genetic and metabolic engineering for the effective removal or detoxification of heavy metals. This book deepens the understanding of microbe-assisted remediation of heavy metal pollutants and provides readers with a comprehensive knowledge of how microbes, along with other advanced strategies, can remediate toxic heavy metals from polluted environments in a sustainable manner. Biotechnological advancements can enhance the efficiency of microbe-based remediation and its possible applications for *in situ* bioremediation of heavy metals.

This book is valuable to a wide range of readers, academicians and researchers, including biotechnologists, environmental engineers, microbiologists, policy makers and students working in related disciplines. We hope this book serves as a resource for academician and researchers for those dedicated in developing innovative, sustainable solutions to one of the major environmental concerns.

Manikant Tripathi
Biotechnology Program, Dr Rammanohar Lohia Avadh University, Ayodhya, Uttar Pradesh, India
manikant.microbio@gmail.com

Sukhminderjit Kaur
Department of Biotechnology, University Institute of Biotechnology, Chandigarh University, Mohali, Punjab, India
sukhminderjit.uibt@cumail.in

Acknowledgements

First and foremost, we express our gratitude to colleagues and teachers for providing us with the support and resources necessary for successfully completing this project. We are especially thankful to our departments and universities for their motivation and the flexibility offered, which was instrumental in bringing this book to its final form. We also thank the Series Editor and Publisher for their critical and insightful feedbacks that helped the overall improvement of this book. We are also thankful to our families, friends and all well-wishers for their constant support in this journey. Lastly, we are thankful to the readers of this book.

Heavy Metal Pollution in Wastewater and its Ecotoxicological Effects: An Overview

1.1 Introduction

Water is absolutely necessary for life on earth, for people as well for the environment, however with increasing pollution day by day due to various anthropogenic activities, freshwater is facing many challenges (Suja *et al.*, 2024). Water pollution is one of the major environmental concerns around the world that have a substantial impact on animals, humans and ecosystems (Yadav *et al.*, 2025). Several pollutants are responsible for water pollution such as organic matter, heavy metals (HMs) and other wastes which are produced by industrial activities like the release of improperly treated wastewater directly into natural water sources (Tripathi *et al.*, 2021; du Plessis, 2022). Water pollution problems are intensified due to a deficiency of regulatory policies and laws, and, where they exist, in implementation, to protect water bodies from mining and industrial wastewater and human actions. Industrial wastewater carries a substantial quantity of toxic HMs that is harmful to aquatic life, humans, plants and the environment (Oladimeji *et al.*, 2024). The US Environmental Protection Agency (EPA) states that in the Philippines, India and Indonesia only 10%, 9% and 14%, respectively, of wastewater is treated in each country (Yadav *et al.*, 2025). HMs are those elements whose atomic number is greater than 20 and atomic density is more than 5 g/cm^2 such as mercury (Hg), cadmium (Cd), chromium (Cr), lead (Pb), arsenic (As), zinc (Zn) and copper (Cu) (Piwowarska *et al.*, 2024). HM pollution has been caused as a result of the progressive development of industrial operations, mining, land filling agricultural water runoff and improper waste management. HMs enter the surroundings through two main types of sources: (i) the natural earth crust and volcanic activity; and (ii) human activities such as municipal sludge, household dust, mining, chemical sectors etc. (Oladimeji *et al.*, 2024). HMs have become a serious problem due to bioaccumulation and their non-biodegradable nature, persisting for a long time in soil and water. Animals and plants consume HMs from various sources, but

Corresponding author: manikant.microbio@gmail.com

© CAB International 2026. *Microbe-assisted Remediation of Heavy Metals in Wastewater* (M. Tripathi and S. Kaur)

1

water is the main one (Rehman *et al.*, 2021). According to the Agency for Toxic Substances and Disease Registry (ATSDR), HMs like Hg, Cd, As, Cr and Pb are highly lethal to both animals and plants (Angon *et al.*, 2024). Metals can have effects on the survival and growth of both aquatic and terrestrial creatures and plants (Zamora-Ledezma *et al.*, 2021).

Heavy metal pollutants in water systems adversely affect living forms in the water. Recent studies have focused on aquatic food product quality that is affected by HM contamination in water habitats; such food products may be consumed by humans and cause negative impacts on their health (Zaynab *et al.*, 2022). Cd and Pb are toxic and can generate many human and animal health concerns including skin sensitivity, genomic instability, nephrotoxicity, different types of cancers, respiratory disease, neurological problems and cardiac diseases (Tripathi *et al.*, 2025). Different diseases are caused by HMs like Hg and Cd in mining workers (Sall *et al.*, 2020).

1.2   Sources of Heavy Metal Toxicity Pollution in Wastewater and Their Biotoxicity

Every year, millions of people are affected by environmental pollution. Current studies have highlighted that various anthropogenic and natural origins are responsible for HM pollution in wastewater (Charkiewicz *et al.*, 2023). The growth of many industrial sectors such as the cloth, leather, paint, battery and glass manufacturing sectors, medicinal sector wastes, household wastes, uses of various type of chemical pesticides and insecticides in agricultures are the main origins of HMs. They may release the HMs directly or via wastewater discharges without proper treatment (Singh and Mishra, 2021). HMs like Pb, As and Cr are discharged by household activity, and industries (Ejaz *et al.*, 2023). Some of the sources of HM pollution in the environment are presented in Figs 1.1 and 1.2.

In India, the use of wastewater for watering crops has resulted in higher HM concentration in vegetables, and has shown notable accumulation leading to health risks such as reduced nutrient intake and hazardous toxicity (Singh *et al.*, 2024a). Mining activities also dispose of highly toxic waste into rivers, posing adverse impacts on aquatic life, particularly to fish populations. HMs can affect the function of organs and produce tissue lesions and tissue damage, which are commonly exploited indicators of pollution (Macklin *et al.*, 2023).

Rivers are also polluted by HMs from agricultural activity, such as high uses of various chemicals containing fertilizers and pesticides for improved crop yields. Singh *et al.* (2024b) studied HMs as one of the contaminants associated with water, while a recent study showed that the world's farming land, lakes and rivers are affected by HMs (Charkiewicz *et al.*, 2023).

1.3   Toxic Impacts on Humans and Animals

The impact of HMs on humans, animals and plants is shown in Fig. 1.3.

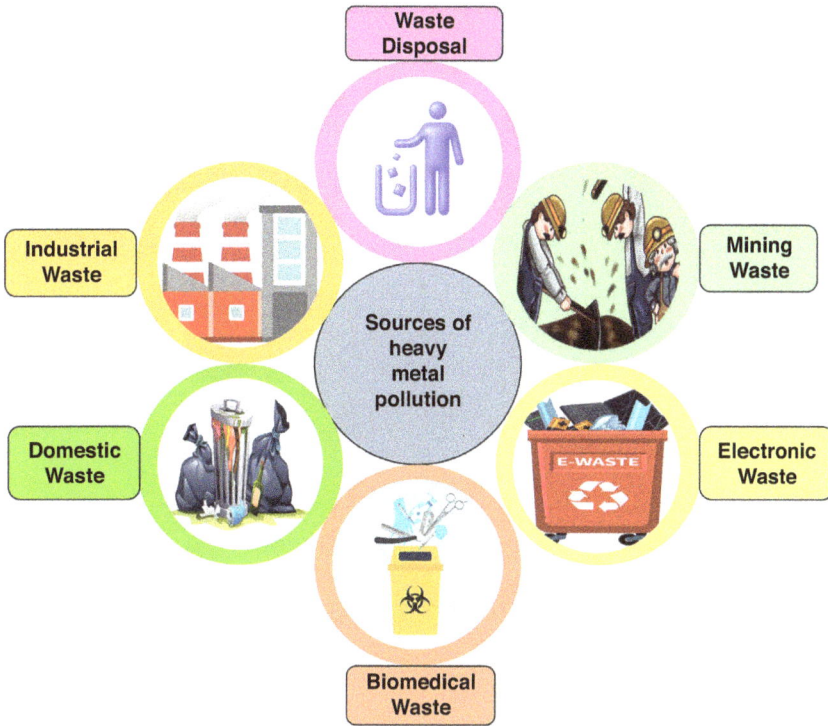

Fig. 1.1. Possible sources of heavy metal pollution in aquatic systems.

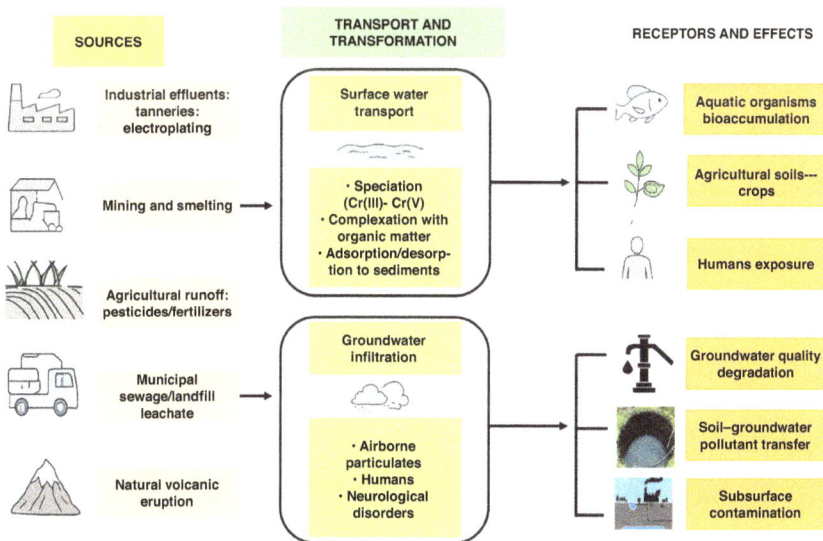

Fig. 1.2. Schematic overview of heavy metal sources, environmental pathways and exposure routes to biota and humans.

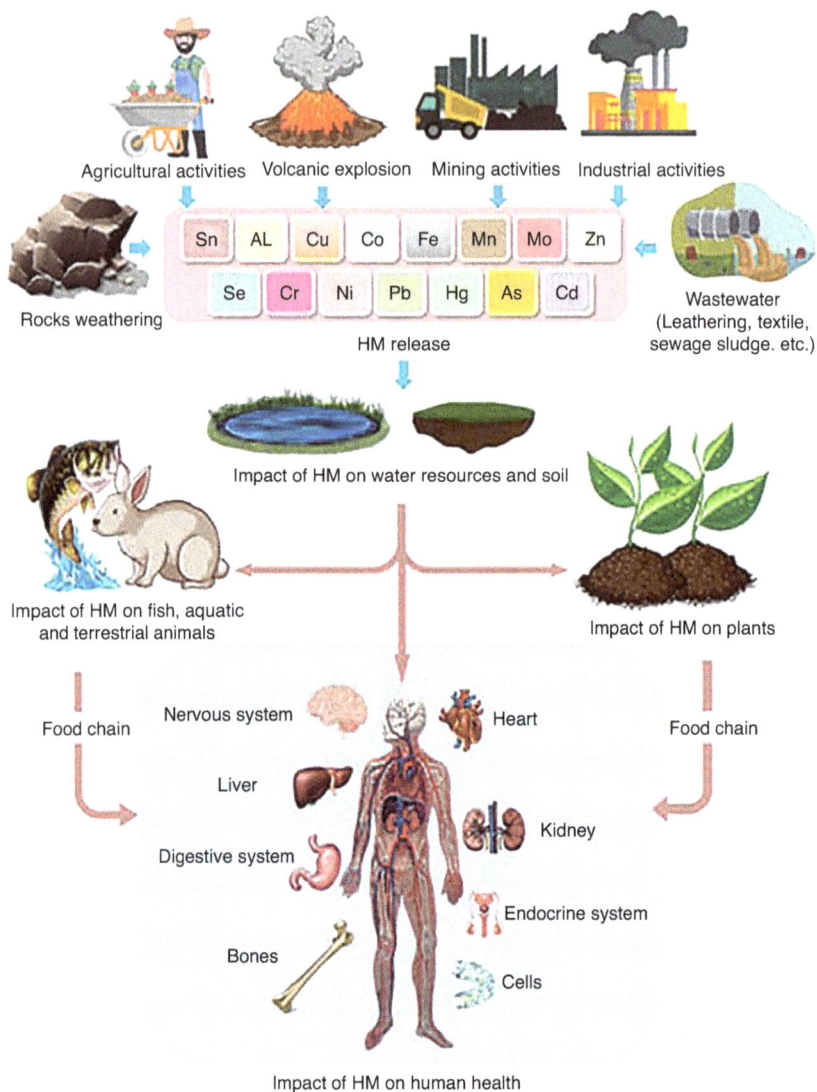

Fig. 1.3. Effect of heavy metals (HMs) on humans, animals and plants. Adapted from Abd Elnabi *et al*. (2023).

1.3.1 Impacts on aquatic life

Wastewater containing HM pollutants, discharged from different sources in freshwater and marine ecosystems, has hazardous impacts on the aquatic flora and fauna. In water systems, fish are easily affected by HM pollution due to HM accumulation in their bodies. Metal affects the immunity system, respiratory system and reproductive organ of fish (Afzaal *et al*., 2022). Water-derived edible products and aquatic plants are important sources of nutrients for humans and thus HM pollutants affect human nutrition directly as well (Sall *et al.*, 2020).

Various types of HM pollution in wastewater and soil are increasing due to ongoing developments like industrialization and urbanization. Balali-Mood *et al.* (2021) discussed the mechanisms of HM toxicity in Hg, Pb, Cr, Cd and As. HMs may inactivate involved enzymes in oxidant stress (Balali-Mood *et al.*, 2021). They cause health problems such as impairment of growth, metabolic abnormalities and cardiac and neurological issues, disrupt hormones causing diabetes and cause dysfunction of the endocrine system and kidney damage (Li *et al.*, 2023). Studies have suggested that a small quantity of manganese (Mn) is required for bone development, the synthesis of developmental hormones, immune and nervous functional activity and energy metabolism in humans and animals. However, an accumulation of high quantities of Mn is toxic; it can promote reactive oxygen species (ROS) formation in rat renal function, leading to genomic damage and oxidative activation (Studer *et al.*, 2022; Yin *et al.*, 2024). In another study, Pan *et al.* (2024) looked at the impact of HMs on the heart, producing cardiovascular diseases like failure of the heart, atherosclerosis, arrhythmia and cardiomyopathy. HMs like Pb, As and Cd have the property of changing heritable DNA sequences due to genetic modification, DNA methylation, etc. (Lamas *et al.*, 2023).

1.3.2 Neurological problems

Heavy metals may affect the proper function of nerve glial cells, microglia, oligodendrocytes and astrocytes. These cells are involved in several important biochemical functional activities concerned with calcium balance, oxidative stress, apoptosis processes and inflammation. Dysfunction of these cells causes neurogenerative disorders like Alzheimer's disease, autism spectrum disorder and Parkinson's disease, and affects many other functions including mitochondrial oxidative function (Ijomone *et al.*, 2025). Autoimmunity is a complex process that originates from both natural and heredity factors. Studies indicate that HMs may impact autoimmune disorders, for instance affecting the central nervous system (Bjørklund *et al.*, 2024). HMs like Pb and Cd accumulate in tissues and cause serious health problems in humans. They have more effect on children than adults, affecting young cardiovascular, kidney and hearing systems (Angon *et al.*, 2024). Excessive amounts of Pb ions dissolved in water are dangerous for humans and animals such as cattle as they cause high blood pressure and may damage the brain (Meftah *et al.*, 2025).

Most industrial, HM-mining countries have increased risks of brain-related illness in humans due to the presence of toxic pollutants that may reduce the production of lipid peroxidation and cause ROS, genomic mutations, modification of proteins, glioma tumorigenesis and brain tumours (Caruso *et al.*, 2023). Researchers have studied the impact of HMs on the proliferation of human tumorigenesis (Hasnain *et al.*, 2024). They have been reported to induce gene modification and to be responsible for cellular dysfunction.

According to researchers, copper may cause gene mutations, whereas nickel and chromium disrupt ion transport channels linked with enzymatic cofactors which may lead to central nervous system injury (Ba *et al.*, 2022).

1.3.3   Toxic impacts on liver and kidney function

Researchers have found that HMs affect renal function. Pb and Cd have a cumulative impact associated with renal tubular abnormality which is greater than the effect of a single metal (Yin *et al.*, 2024). Chronic exposure to Cd has also been linked to kidney and liver damage (Yin *et al.*, 2024). Cd is a notable harmful metal mostly linked to nephron-damaging issues; it promotes DNA lesions and oxidative stress and dysfunction of the renal tubules and nephrons (Wang *et al.*, 2021). Pb generally originates from industrial waste and has harmful impacts on human health leading to the formation of ROS, which may be responsible for epithelial cell death and a reduction in the activity of kidneys. Pb causes kidney damage via the inhibition of calcium homeostasis, which induces inflammation and has negative effects on the heart and nervous systems (Ma *et al.*, 2022; Yin *et al.*, 2024). Small quantities of Cd can cause serious diseases like leukaemia, genetic disorders, emphysema, bronchitis and asthma. HMs can induce lung swelling via oxidative stress, resulting in lung tissue damage (Charkiewicz *et al.*, 2023).

Studies also indicate that HMs affect hepatic function. Increasing concentration of HMs enter the liver via contaminated food and water, causing toxicity in the liver (Li *et al.*, 2023). Metabolic dysfunction-associated fatty liver disease (MAFLD) in adults is a common chronic hepatotoxicity disorder worldwide. Non-alcoholic fatty liver disease (NAFLD) is produced from inactivation of fat metabolites and higher concentrations of selenium (Se) and Mn, and affects MAFLD (Tang *et al.*, 2024).

1.3.4   Endocrine disruption

Recent investigations by researchers indicate that some HMs play an important role in hormonal systems, leading to developmental health complications (Liu *et al.*, 2023). Pb toxicity can cause endocrine system disruption in humans. Pb is released by various sectors (mainly from petrol, the alloy industry and glazes) and enters human cells, where its toxicity disrupts levels of pituitary and developmental hormones (Li *et al.*, 2023). Pb toxicity infects the body's immunity system, affecting cytokines and tumour necrosis factor (TNF) cells; the disruption of these constrains heme formation enzymes like ferrochelate, δ-aminolevulinic acid dehydratase and glutathione, reducing the synthesis of heme and causing anaemia (Balali-Mood *et al.*, 2021). Table 1.1 shows different sources of HM pollution and its adverse health effects on humans.

Table 1.1. Heavy metal pollution sources and their toxic impacts on animals and humans.

Heavy metal	Sources	Impacts on human health	Reference
Nickel (Ni)	Nickel alloy generation and stainless-steel material	Lung fibrosis, skin disease, kidney	Kapepula and Luis, 2024
Cadmium (Cd)	Electroplating, insecticides, synthetic chemical industries	Liver, lung, cancer, pulmonary and developmental issues. ROS production	Zamora-Ledezma *et al.*, 2021
Mercury (Hg)	Seafood, fish, industrial wastewater, pharmaceutical sectors, cosmetic sectors	CNS damage, neurotoxicity, kidney dysfunction, ulcers, hepatotoxicity	Qosem *et al.*, 2021; Pan *et al.*, 2024
Lead (Pb)	Lead-mediated batteries, alloys, glazes, cable-sheathing pigments	Immunity system, anaemia, increased inflammatory cytokines	Balali-Mood *et al.*, 2021
Arsenic (As)	Forming waste, mining, metal sectors, nature	Liver dysfunction, dermal and hair issues, heart disease	Balali-Mood *et al.*, 2021
Chromium (Cr)	Tanneries, pulp mills and steel industries, textile industries, metal plating	Pancreas, lungs, skin disease, kidney, genetic instability	Angon *et al.*, 2024

1.4   Toxic Impacts on Plants

Researchers have studied the effects of HM toxicity on plants and the different ways HMs enter the ecosystem, such as point and non-point origins. HMs can accumulate in soils, affecting plant growth. Currently, the agricultural sector has received considerable attention due to the harmful effects of HMs that directly affect plant developments, biological activity and productivity (Pande *et al.*, 2022). High concentrations of As, Cr, Pb and Hg are very harmful to plant life. Toxic metals decrease the transport of nutrients and water in plants and inhibit plant growth by increasing oxidative stress; they inhibit the germination of seeds, plant growth, nutrient quality and important cellular functions and can cause chlorosis. Cd affects physio-biochemical activity in plants and as a result the growth of plants is reduced (Ali and Gill, 2022). Cr occurs naturally in both trivalent Cr^{+3} and hexavalent Cr^{+6} forms in environment. Cr^{+6} is toxic to the ecosystem (Tripathi *et al.*, 2025), and its toxicity affects the growth of plants as well as the germination of seeds, photosynthetic rate, oxidative stress and enzymatic activity (Ma *et al.*, 2022).

The metals can be transported into plants via various routes such as symbiotic, apoplectic, root, xylem and phloem pathways (Vasilachi *et al.*, 2023). Nickel toxicity impacts the seed germination and growth of many plants species by decreasing the enzyme activity necessary for germination (Singh, 2020). Copper and lead pollution has a significant impact on the antioxidants of two species of mangrove plants, *Avicennia alba* and *Excoecaria agallocha* (Rozirwan *et al.*, 2025).

The effect of HMs on various plants are presented in Table 1.2. These studies suggest that HM discharge or amendment in the soil poses negative effects on plant growth. These toxic metals enter plant parts and adversely affect their physiological and biochemical functions, resulting in decreased growth of plants. Careful monitoring of these toxicants and their appropriate mitigation from contaminated soils are necessary steps for possible agriculture and environmental sustainability.

1.4.1 Effect on photosynthesis rate in plants

Heavy metals may enter plant cells and reduce the chlorophyll content and light energy efficiency, resulting in a decrease in the photosynthesis rate. Cu and Pb are generated by mining and industrial activities, municipal sludge, agricultural pesticides and natural sources. Micronutrient Cu is a vital component for the physiological and biochemical activity of plants but high concentrations of Cu and Pb cause harm to plant physiological activity (Giannakoula *et al.*, 2021). The presence of HM pollutant in soil has a negative impact on

Table 1.2. The harmful effects of heavy metals on different plant species.

Heavy metal	Plants	Impacts on plant health	Reference
Lead (Pb)	*Spinacea oleracea* L.	Decreases dry weight potassium levels in shoots and roots	Kibria *et al.*, 2010
Mercury (Hg)	Tomato (*Lycopersicon esculentum* Mill.)	Reduces germination, survival percentage, plant height, root length and fruit weight	Shekar *et al.*, 2011
Cadmium (Cd)	Garlic (*Allium sativum*)	Reduces shoot elongation	Jiang *et al.*, 2001; Anas *et al.*, 2025
Cadmium (Cd)	Rice (*Oryza sativa*)	Affects spikelet fertility and decreases productivity	Xia *et al.*, 2024
Arsenic (As), lead (Pb)	Wheat (*Triticum* sp.)	Reduces growth, biomass and seed germination	Anas *et al.*, 2024
Copper (Cu), zinc (Zn), chromium (Cr)	Paddy plants	Affects roots, shoots, seed germination and zinc finger proteins dysfunction	Angon *et al.*, 2024

agricultural food security. HMs, even in small quantities, can affect paddy crop plants, and Cd pollutants reduce the edible quality of rice (Xia *et al.*, 2024).

1.5 Challenges and Future Prospects

The pollution of wastewater with HMs like arsenic, chromium, lead, cadmium, mercury, etc. is deleterious to the ecosystem because of the toxic impacts and harmful effects of HMs on the health of humans, animals and plants. These toxic HMs originate from point and non-point sources. Increasing concentrations of toxic HMs in the food chain has resulted in significant adverse health impacts to the ecosystem. These toxicants also contaminate drinking water sources. HMs exist in several oxidation states and toxic forms may biotransform into non-toxic or less harmful forms.

Proper regulatory guidelines and monitoring systems are needed from environmental government and non-government regulatory authorities to manage HM pollution from wastewater. Currently, industrial or domestic wastewater may be released into the environment without proper monitoring of toxic pollutants, but should be properly treated before discharge into the aquatic ecosystem. Understanding the ecotoxicological impacts of HMs is a challenge that requires sustained efforts. There is also a need for environmentally friendly strategies to detoxify these HMs in contaminated ecosystems. Advanced wastewater treatment strategies using nanotechnology, microbial bioremediation and integrated technologies would be helpful in the detoxification of HMs.

1.6 Conclusion

Heavy metal toxicity in aquatic systems is a major environmental concern. An increasing world population and rise in industrial and human activities, producing more industrial and household waste and agricultural runoff, are the main sources of toxic HM pollution in wastewater. Improperly or untreated wastewater directly released into natural water sources may cause pollution in aquatic systems and pose adverse effects to all ecosystems. HM pollutants have harmful effects on aquatic life forms, humans and plants. It should be a top priority to have inclusive policies to restrict HM pollution in wastewater. This chapter has indicated the necessity of regulation policies for industrial wastewater, with proper monitoring and treatment before its release into ecosystems as these wastewaters may contain various toxic HMs. Green phytoremediation and other environmentally friendly tools may be used to manage HM pollution in water and, ultimately, for safer and sustainable ecosystems.

References

Abd Elnabi, M.K., Elkaliny, N.E., Elyazied, M.M., Azab, S.H., Elkhalifa, S.A. *et al.* (2023) Toxicity of heavy metals and recent advances in their removal: A review. *Toxics* 11(7), 580. DOI: 10.3390/toxics11070580.

Afzaal, M., Hameed, S., Liaqat, I., Ali Khan, A.A., Abdul Manan, H. *et al.* (2022) Heavy metals contamination in water, sediments and fish of freshwater ecosystems in Pakistan. *Water Practice and Technology* 17(5), 1253–1272. DOI: 10.2166/wpt.2022.039.

Ali, B. and Gill, R.A. (2022) Editorial: Heavy metal toxicity in plants: Recent insights on physiological and molecular aspects. *Frontiers in Plant Science* 12, 830682. DOI: 10.3389/fpls.2021.830682.

Anas, M., Saeed, M., Naeem, K., Shafique, M.A. and Quraishi, U.M. (2024) Anatomical and ionomics investigation of bread wheat (*Triticum aestivum* L.) to decipher tolerance mechanisms under arsenic stress. *Journal of Plant Growth Regulation* 43(10), 3609–3625. DOI: 10.1007/s00344-024-11332-9.

Anas, M., Khattak, W.A., Fahad, S., Alrawiq, N., Alrawiq, H.S. *et al.* (2025) Mitigating heavy metal pollution in agriculture: A multi-omics and nanotechnology approach to safeguard global wheat production. *Journal of Hazardous Materials Advances* 17, 100584. DOI: 10.1016/j.hazadv.2024.100584.

Angon, P.B., Islam, M.S., Kc, S., Das, A., Anjum, N. *et al.* (2024) Sources, effects and present perspectives of heavy metals contamination: Soil, plants and human food chain. *Heliyon* 10(7), e28357. DOI: 10.1016/j.heliyon.2024.e28357.

Ba, Q., Zhou, J., Li, J., Cheng, S., Zhang, X. *et al.* (2022) Mutagenic characteristics of six heavy metals in *Escherichia coli*: The commonality and specificity. *Environmental Science & Technology* 56(19), 13867–13877. DOI: 10.1021/acs.est.2c04785.

Bjørklund, G., Đorđević, A.B., Hamdan, H., Wallace, D.R. and Peana, M. (2024) Metal-induced autoimmunity in neurological disorders: A review of current understanding and future directions. *Autoimmunity Reviews* 23(3), 103509. DOI: 10.1016/j.autrev.2023.103509.

Balali-Mood, M., Naseri, K., Tahergorabi, Z., Khazdair, M.R. and Sadeghi, M. (2021) Toxic mechanisms of five heavy metals: Mercury, lead, chromium, cadmium, and arsenic. *Frontiers in Pharmacology* 12, 643972. DOI: 10.3389/fphar.2021.643972.

Caruso, G., Nanni, A., Curcio, A., Lombardi, G., Somma, T. *et al.* (2023) Impact of heavy metals on *Glioma tumorigenesis*. *International Journal of Molecular Sciences* 24(20), 15432. DOI: 10.3390/ijms242015432.

Charkiewicz, A.E., Omeljaniuk, W.J., Nowak, K., Garley, M. and Nikliński, J. (2023) Cadmium toxicity and health effects-a brief summary. *Molecules (Basel, Switzerland)* 28(18), 6620. DOI: 10.3390/molecules28186620.

du Plessis, A. (2022) Persistent degradation: Global water quality challenges and required actions. *One Earth* 5(2), 129–131. DOI: 10.1016/j.oneear.2022.01.005.

Ejaz, U., Khan, S.M., Khalid, N., Ahmad, Z., Jehangir, S. *et al.* (2023) Detoxifying the heavy metals: A multipronged study of tolerance strategies against heavy metals toxicity in plants. *Frontiers in Plant Science* 14, 1154571. DOI: 10.3389/fpls.2023.1154571.

Giannakoula, A., Therios, I. and Chatzissavvidis, C. (2021) Effect of lead and copper on photosynthetic apparatus in citrus (*Citrus aurantium* L.) plants. The role of antioxidants in oxidative damage as a response to heavy metal stress. *Plants (Basel, Switzerland)* 10(1), 155. DOI: 10.3390/plants10010155.

Hasnain, M., Qayoom, A., Saira, A., Hussain, M., Rehman, A. *et al.* (2024) Implications of heavy metals on human cancers. *OALib* 11(12), 1–23. DOI: 10.4236/oalib.1112644.

Ijomone, O.K., Ukwubile, I.I., Aneke, V.O., Olajide, T.S., Inyang, H.O. *et al.* (2025) Glial perturbation in metal neurotoxicity: Implications for brain disorders. *Neuroglia* 6(1), 4. DOI: 10.3390/neuroglia6010004.

Jiang, W., Liu, D. and Hou, W. (2001) Hyperaccumulation of cadmium by roots, bulbs and shoots of garlic (*Allium sativum* L.). *Bioresource Technology* 76(1), 9–13. DOI: 10.1016/s0960-8524(00)00086-9.

Kapepula, V.L. and Luis, P. (2024) Removal of heavy metals from wastewater using reverse osmosis. *Frontiers in Chemical Engineering* 6, 1334816. DOI: 10.3389/fceng.2024.1334816.

Kibria, M., Maniruzzaman, M., Islam, M. and Osman, K. (2010) Effects of soil applied lead on growth and partitioning of ion concentration in *Spinacea oleracea* L. tissues. *Soil and Environment* 29(1), 1–6.

Lamas, G.A., Bhatnagar, A., Jones, M.R., Mann, K.K., Nasir, K. *et al.* (2023) Contaminant metals as cardiovascular risk factors: A scientific statement from the american heart association. *Journal of the American Heart Association* 12(13), e029852. DOI: 10.1161/JAHA.123.029852.

Liu, D., Shi, Q., Liu, C., Sun, Q. and Zeng, X. (2023) Effects of endocrine-disrupting heavy metals on human health. *Toxics* 11(4), 322. DOI: 10.3390/toxics11040322.

Ma, J., Alshaya, H., Okla, M.K., Alwasel, Y.A., Chen, F. *et al.* (2022) Application of cerium dioxide nanoparticles and chromium-resistant bacteria reduced chromium toxicity in sunflower plants. *Frontiers in Plant Science* 13, 876119. DOI: 10.3389/fpls.2022.876119.

Li, Y., Ortiz, R.G.G., Uyen, P.T.M., Cong, P.T., Othman, S.I. *et al.* (2023) Environmental impact of endocrine-disrupting chemicals and heavy metals in biological samples of petrochemical industry workers with perspective management. *Environmental Research* 231(Part 2), 115913. DOI: 10.1016/j.envres.2023.115913.

Macklin, M.G., Thomas, C.J., Mudbhatkal, A., Brewer, P.A., Hudson-Edwards, K.A. *et al.* (2023) Impacts of metal mining on river systems: A global assessment. *Science (New York, N.Y.)* 381(6664), 1345–1350. DOI: 10.1126/science.adg6704.

Meftah, S., Meftah, K., Drissi, M., Radah, I., Malous, K. *et al.* (2025) Heavy metal polluted water: Effects and sustainable treatment solutions using bio-adsorbents aligned with the sdgs. *Discover Sustainability* 6(1), 137. DOI: 10.1007/s43621-025-00895-6.

Oladimeji, T.E., Oyedemi, M., Emetere, M.E., Agboola, O., Adeoye, J.B. *et al.* (2024) Review on the impact of heavy metals from industrial wastewater effluent and removal technologies. *Heliyon* 10(23), e40370. DOI: 10.1016/j.heliyon.2024.e40370.

Pan, Z., Gong, T. and Liang, P. (2024) Heavy metal exposure and cardiovascular disease. *Circulation Research* 134(9), 1160–1178. DOI: 10.1161/CIRCRESAHA.123.323617.

Pande, A., Mun, B.-G., Methela, N.J., Rahim, W., Lee, D.-S. *et al.* (2022) Heavy metal toxicity in plants and the potential NO-releasing novel techniques as the impending mitigation alternatives. *Frontiers in Plant Science* 13, 1019647. DOI: 10.3389/fpls.2022.1019647.

Piwowarska, D., Kiedrzyńska, E. and Jaszczyszyn, K. (2024) A global perspective on the nature and fate of heavy metals polluting water ecosystems, and their impact and remediation. *Critical Reviews in Environmental Science and Technology* 54(19), 1436–1458. DOI: 10.1080/10643389.2024.2317112.

Qosem, A.A., Mohammed, R.H. and Lawal, D.U. (2021) Removal of heavy metal ions from wastewater: A comprehensive and critical review. *Npj Clean Water* 4, 36. DOI: 10.1038/s41545-021-00127-0.

Rehman, A.U., Nazir, S., Irshad, R., Tahir, K., ur Rehman, K. *et al.* (2021) Toxicity of heavy metals in plants and animals and their uptake by magnetic iron oxide nanoparticles. *Journal of Molecular Liquids* 321, 114455. DOI: 10.1016/j.molliq.2020.114455.

Rozirwan, R., Khotimah, N.N., Putri, W.A.E., Fauziyah, F., Aryawati, R. *et al.* (2025) Biomarkers of heavy metals pollution in mangrove ecosystems: Comparative assessment in industrial impact and conservation zones. *Toxicology Reports* 14, 102011. DOI: 10.1016/j.toxrep.2025.102011.

Sall, M.L., Diaw, A.K.D., Gningue-Sall, D., Efremova Aaron, S. and Aaron, J.-J. (2020) Toxic heavy metals: Impact on the environment and human health, and treatment with conducting organic polymers, a review. *Environmental Science and Pollution Research International* 27(24), 29927–29942. DOI: 10.1007/s11356-020-09354-3.

Shekar, C.H.C., Sammaiah, D., Shasthree, T. and Reddy, K.J. (2011) Effect of mercury on tomato growth and yield attributes. *International Journal of Pharma and Bio Sciences* 2(2), B-358–B364.

Singh, P. (2020) The effect of nickel on seed germination and seeding growth of *Raphanus sativus* cv. Jaunpuri. *Journal of Emerging Technologies and Innovative Research (JETIR)* 7(8), 1328–1330. DOI: 10.56975/jetir.v7i8.551284.

Singh, V. and Mishra, V. (2021) Microbial removal of cr (VI) by a new bacterial strain isolated from the site contaminated with coal mine effluents. *Journal of Environmental Chemical Engineering* 9(5), 106279. DOI: 10.1016/j.jece.2021.106279.

Singh, V., Ahmed, G., Vedika, S., Kumar, P., Chaturvedi, S.K. *et al.* (2024b) Toxic heavy metal ions contamination in water and their sustainable reduction by eco-friendly methods: Isotherms, thermodynamics and kinetics study. *Scientific Reports* 14(1), 7595. DOI: 10.1038/s41598-024-58061-3.

Singh, R., Gupta, S., Khare, A.K. and Tiwari, S. (2024a) Heavy metal contamination through wastewater irrigation on the soil and vegetables: Impact on the nutrient content and health risks. *Crop Research* 59(1 and 2), 52–59. DOI: 10.31830/2454-1761.2024.CR-944.

Studer, J.M., Schweer, W.P., Gabler, N.K. and Ross, J.W. (2022) Functions of manganese in reproduction. *Animal Reproduction Science* 238, 106924. DOI: 10.1016/j.anireprosci.2022.106924.

Suja, S.K., Almaas, S., Gracy, A.P., Gowsika, P., Jeyapradeepa, K. *et al.* (2024) Contamination of water by heavy metals and treatment methods – a review. *Current World Environment* 19(1), 04–21. DOI: 10.12944/CWE.19.1.2.

Tang, S., Luo, S., Wu, Z. and Su, J. (2024) Association between blood heavy metal exposure levels and risk of metabolic dysfunction associated fatty liver disease in adults: 2015-2020 NHANES large cross-sectional study. *Frontiers in Public Health* 12, 1280163. DOI: 10.3389/fpubh.2024.1280163.

Tripathi, M., Kumar, S., Yadav, S.K. and Prasad, N. (2021) Heavy metal pollution in Indian river and its impact on health. In: *Environmental Communication Lab to Land*. India: Shree Publisher and Distributor, pp. 94–105.

Tripathi, M., Singh, R., Lal, B., Haque, S., Ahmad, I. *et al.* (2025) Marine microbial bioremediation of heavy metal contaminants in waste water for health and environmental sustainability: A review. *Indian Journal of Microbiology* 65(2), 573–582. DOI: 10.1007/s12088-024-01427-y.

Vasilachi, I.C., Stoleru, V. and Gavrilescu, M. (2023) Analysis of heavy metal impacts on cereal crop growth and development in contaminated soils. *Agriculture* 13(10), 1983. DOI: 10.3390/agriculture13101983.

Wang, M., Chen, Z., Song, W., Hong, D., Huang, L. *et al.* (2021) A review on cadmium exposure in the population and intervention strategies against cadmium toxicity. *Bulletin of Environmental Contamination and Toxicology* 106(1), 65–74. DOI: 10.1007/s00128-020-03088-1.

Xia, W., Ghouri, F., Zhong, M., Bukhari, S.A.H., Ali, S. *et al.* (2024) Rice and heavy metals: A review of cadmium impact and potential remediation techniques. *Science of the Total Environment* 957, 177403. DOI: 10.1016/j.scitotenv.2024.177403.

Yadav, V., Manjhi, A. and Vadakedath, N. (2025) Assessment of heavy metals content and diversity of mercury-tolerant bacteria in heavy metal-polluted environmental samples and mercury bioremediation ability of isolated strains. *Environmental Advances* 19, 100624. DOI: 10.1016/j.envadv.2025.100624.

Yin, G., Zhao, S., Zhao, M., Xu, J., Ge, X. *et al.* (2024) Complex interplay of heavy metals and renal injury: New perspectives from longitudinal epidemiological evidence. *Ecotoxicology and Environmental Safety* 278, 116424. DOI: 10.1016/j.ecoenv.2024.116424.

Zamora-Ledezma, C., Negrete-Bolagay, D., Figueroa, F., Zamora-Ledezma, E., Ni, M. *et al.* (2021) Heavy metal water pollution: A fresh look about hazards, novel and conventional remediation methods. *Environmental Technology & Innovation* 22, 101504. DOI: 10.1016/j.eti.2021.101504.

Zaynab, M., Al-Yahyai, R., Ameen, A., Sharif, Y., Ali, L. *et al.* (2022) Health and environmental effects of heavy metals. *Journal of King Saud University – Science* 34(1), 101653. DOI: 10.1016/j.jksus.2021.101653.

Bacteria-based Methods for Remediation of Heavy Metals from Polluted Sites: Recent Updates

2.1 Introduction

The contamination of heavy metals (HMs) has emerged as one of the most critical problems in the environment across the world due to their inability to be dissolved in water, their non-biodegradability and their bioaccumulative quality. There is also a high concentration of toxic metals such as cadmium (Cd), lead (Pb), mercury (Hg), chromium (Cr), arsenic (As), nickel (Ni), copper (Cu) and zinc (Zn) in soil- and water-based systems due to rapid industrialization, urbanization and anthropogenic activities (e.g. mining, electroplating, battery production and improper disposal of e-waste) (Briffa *et al.*, 2020). These metals are toxic to microorganisms as well as dangerous to health as their presence disrupts essential homeostatic functions, causes oxidative stress and harms different proteins and cellular membranes (Jan *et al.*, 2015). HMs inhibit growth in plants, result in chlorosis and lower crop yields because they affect photosynthesis and the absorption of nutrients (Ilyas *et al.*, 2025). Therefore, the reduction of HM contamination is crucial in terms of preserving the environment and reducing health risks to people. Chemical precipitation, membrane filtration, reverse osmosis, ion exchange and electrochemical treatments are among the traditional remediation technologies broadly employed to treat HMs in contaminated environments. These methods, however, are costly, energy-inefficient and typically create secondary waste that makes them less feasible in terms of scale and sustainability (Shofia *et al.*, 2025).

In comparison, bioremediation, which is the use of living organisms or their metabolic products, is a cost-effective, environmentally friendly and sustainable way to detoxify sites polluted with HMs (Pham *et al.*, 2022). Moreover, bacteria – when used in bioremediation techniques – are useful because of their high adaption capabilities and rapid growth, as well as their ability to survive under diverse, harsh conditions. Bacteria use various mechanisms to reduce the bioavailability and toxicity of HMs in contaminated

Corresponding author: sukhminderjit.uibt@cumail.in

environments. Biosorption, bioaccumulation and efflux systems are the three most important mechanisms studied in bacteria that work together to mitigate metal stress. Biosorption is an energy-independent process where metals ions are exchanged for functional groups (carboxyl, phosphate, amino and hydroxyl) of bacterial cell walls and extracellular polymeric substance (EPS) through ion exchange, electrostatic interaction and complexation without metabolism (Guo *et al.*, 2020). This is a rapid, reversible process and effective at low metal concentrations; it is frequently studied as a treatment mechanism in wastewater. By contrast, in bioaccumulation HMs are mobilized into the cytoplasm of the bacteria, where cells actively consume energy to adsorb metal ions. The HMs are then detoxified by binding to metal-binding proteins like metallothioneins or peptides like glutathione (Torres, 2020). Biosorption allows bacteria to survive in poisonous conditions and is also essential to sequestrate metal over time. Another important defense mechanism involves an efflux system, which actively removes the cytoplasmic toxic metal ions, aided by localized transporters: P-type ATPases, cation diffusion facilitators (CDFs) and resistance–nodulation–division (RND) systems (Kim *et al.*, 2019). The pumps control intracellular metal homeostasis and prevent increases in metal concentration in the cell. These mechanisms work together to support the survival of microbes along with their increased bioremediation efficiency. In this context, recent developments in the field of molecular biology, genetic engineering and nanotechnology have enhanced bacterial processes related to metal remediation even further. One method is through CRISPR (clustered regularly interspaced short palindromic repeats)-Cas-based genome editing to produce a higher number of metal-binding and more tolerant bacteria (Karnwal *et al.*, 2025). It has also been demonstrated that nanoparticles and biofilms lead to better metal recovery and metal recyclability (Gol-Soltani *et al.*, 2024). These advances mean that bacteria-based treatments should be given greater priority as alternatives to traditional remediation solutions.

This chapter gives a detailed overview of the basic mechanisms used by bacteria for HM detoxification such as biosorption, bioaccumulation and efflux systems. It also elaborates the molecular basis of the various HM removal processes, their ecological significance and possible applications in the current bioremediation technology.

2.2   Different Sources of Heavy Metal Pollution in Wastewater

Urbanization and industrial activities are the biggest source of HMs discharged into the environment (Fig. 2.1). There are many industries that release HMs in their effluent directly into water bodies such as rivers. Wastewater from industries such as mining, metal processing, electroplating, battery manufacturing, textile, dyes, leather tanning, paper, plastics, petrochemicals and electronics is a leading source. These activities release metals like Cr, Pb, Hg, Cd, Ni, Cu, Zn

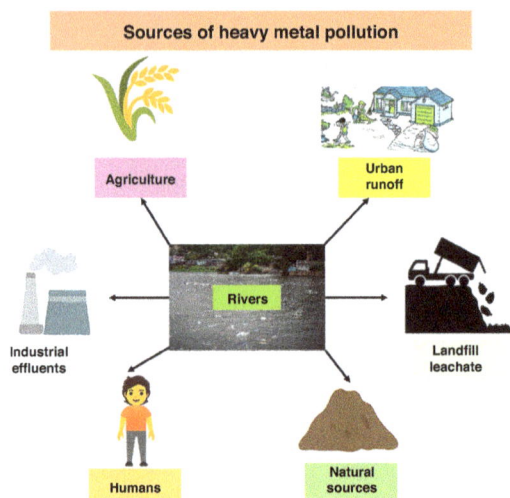

Fig. 2.1. Different sources of heavy metal pollution in wastewater.

and As into water bodies. The recycling of batteries and electronic devices contribute Pb, Cd, Ni and Hg to wastewater stream (Nnaji *et al.*, 2023). Another prominent contributor are the processes of mining and smelting operations. Discharges from mining for metals or coal, as well as smelting and refining metal ores, result in wastewater containing high concentrations of various HMs such as Pb, As, Cd, Cr and Cu (Aziz *et al.*, 2023). Weathering of rocks, volcanic eruptions, forest fires and soil erosion also release metals, though these are generally less significant compared with human activities (Dagdag *et al.*, 2023). Various agriculture practices such as the use of fertilizers, pesticides, herbicides and sewage sludge in agriculture introduces metals including As, Cd, Cu and Zn into surface water and groundwater. Leachate from landfills and sewage systems may carry HMs from household chemicals, piping, paints and personal care products. Rainwater runoff from roads and urban surfaces can collect metals like Pb, Cu and Zn from vehicle emissions, brakes and roofing material. Human activities are also responsible for the deposition of Cd by manufacturing processes like metal ore combustion, the use of fossil fuels and waste burning. The use of paints, preservatives, insecticides and phosphate fertilizers, etc., also emit HMs and they too accumulate in ecosystems causing serious threat to the environment (Nnaji *et al.*, 2023).

2.3 Heavy Metal Toxicity

Heavy metals present a major risk to human health, animal populations and plant ecosystems due to their ability to disrupt metabolic processes and mimic or interfere with essential biological functions. Some metals are eliminated, while others bioaccumulate, causing chronic effects through oxidative stress and free radical damage. The severity of toxicity depends on dose, exposure route and duration. HMs such as Hg, Pb, Cd, Cr and As cause toxic effects through mechanisms like oxidative stress, enzyme inhibition and DNA damage. Their bioaccumulation disrupts critical cellular functions and contributes to carcinogenesis. Despite known hazards, poisoning remains prevalent, highlighting the need for preventive strategies and improved treatments (Zahran *et al.*, 2025).

In humans, exposure to HMs can lead to damage to various organs. Chronic exposure of HMs can lead to an accumulation of these metals in the body that disrupts the normal function of critical organs, including the kidneys, liver, lungs, heart and brain. They can lead to neurodevelopmental and cognitive disorders, particularly in children. Symptoms range from fatigue, anxiety and memory loss to intellectual impairment and diseases like Alzheimer's and Parkinson's disease (Jomova *et al.*, 2025). Some metals (As, Cd, Cr) are established human carcinogens, disrupting DNA synthesis and repair, and driving cancer development. HMs can alter immune function and hormone regulation, increasing susceptibility to infections and metabolic disorders, skin lesions, vascular damage, birth defects, reproductive toxicity and gastrointestinal issues, reflecting the systemic toxicity of metals (Jaishankar *et al.*, 2014).

HMs also harm other animals like fish, birds and mammals. In animals, HMs can accumulate in their tissues, leading to organ toxicity. Similar to humans, animals suffer from kidney, liver, nervous system and cardiovascular impairment due to chronic exposure. Metals can disrupt the neural and endocrine systems of animals, subsequently affecting their behaviour and growth resulting in reduced fertility, offspring deformities and increased mortality (Balali-Mood *et al.*, 2021).

In plants, HM toxicity interferes with plant metabolism, inhibits root and shoot development and reduces chlorophyll content, resulting in stunted growth and reduced crop yield. Metals cause cellular damage in plants by generating reactive oxygen species (ROS), compromising cell membranes and enzymes, leading to oxidative stress. Root exposure to metals can disrupt nutrient absorption, causing deficiencies and overall plant decline. Accumulated metals in plant tissues transfer to herbivores and up the food chain, exacerbating systemic toxicity (Balali-Mood *et al.*, 2021).

2.4   Heavy Metal Remediation

Environmental pollution due to HM particulates from industries is increasing day by day and they are entering food chains and ultimately reaching animals and humans. Various conventional methods have been used for HM remediation such as chemical precipitation, electrodialysis and electrodialysis reversal, membrane filtration, ultrafiltration, nanofiltration, reverse osmosis, microfiltration and photocatalysis. These conventional techniques are expensive and lack environmentally friendly solutions. Bioremediation, a technique utilizing living organisms to remove HMs offers an economical approach. Bioremediation uses both live microorganisms as well as plants (phytoremediation) for the removal of HMs (Pham *et al.*, 2022). The nature of microorganisms mainly relies on the type of polluting agent. Bioremediation can be achieved by mainly two processes – either naturally (*in situ*) or through intervention processes (*ex situ*).

In situ bioremediation involves onsite treatment of the contaminated effluent. The HMs are directly removed using different microorganisms at the site of the contamination. *In situ* bioremediation is of two types: intrinsic *in situ* bioremediation and engineered *in situ* remediation. Naturally occurring microorganisms in the contaminated site are stimulated to remove HMs in the former process. Engineered *in situ* remediation is where the systems are incorporated in the existing conditions in order to enhance the microflora at the site (Vidali, 2001; Evans and Furlong, 2003). This technique is used for the bioremediation of various pollutants such as dyes, HMs, chlorinated solvents, etc. (Roy *et al.*, 2015). The success of *in situ* methods of bioremediation depend on various factors such as moisture content, pH, temperature, nutrient availability, status of electron acceptors, etc. (Atlas and Philp, 2005). Since the method uses naturally occurring microorganisms it is cheaper as well as harmless, nor does it produces harmful byproducts in the environment, with no disturbance to its structure. The other advantage of *in situ* bioremediation is that it involves minimal exposure of public and site personnel.

Ex situ bioremediation, as the name suggests, involves the removal of the contaminated effluent so it can be treated elsewhere.

2.5   Mechanisms of Heavy Metal Degradation by Bacteria

Heavy metal contamination of the environment presents an extreme eco-logical and health risk because of the persistence, non-biodegradability and bioaccumulative properties of the elements involved. Bacteria are the most commonly employed microorganisms for removing HMs from polluted sites. Bacteria are single-celled prokaryotic microorganisms that can thrive in a wide range of environmental settings due to their small size, rapid growth rate and easy cultivation (Tarfeen *et al.*, 2022). Bacteria use various mechanisms to decrease the bioavailability of HMs in the environment. They can interact with HMs through direct or indirect mechanisms depending on the nature of the bacteria, the type of metal and the surrounding environment. Various factors, such as temperature, pH, nutrient sources and metal ions, are important for governing the bioavailability and mobility of HMs for transformation processes by microorganisms. HMs may disrupt microbial cell membranes, but bacteria possess characteristic enzymatic profiles that can overcome toxic effects. Usually, HMs attach to functional groups such as the amino, carboxyl, sulfate and phosphate groups present on the polysaccharide layers of the bacterial cell wall (Yue *et al.*, 2015; Yin *et al.*, 2016). Bacteria hold a high potential for the removal of HMs from contaminated environments using different mechanisms. In response to metal toxicity, bacteria have developed various mechanisms to survive and to detoxify in conditions contaminated with metal. Of these, biosorption, bioaccumulation and efflux systems are central in reducing the toxicity of the HMs (Karnwal *et al.*, 2025). Biosorption is an important metabolism-independent and passive process that depends upon the reaction of metallic ions with the functional groups on the cell wall of the microbe; bioaccumulation is an active uptake and internal storage of metals. On the other hand, efflux systems represent an energy-dependent transport system involved in the expulsion of metal ions, hence intracellular homeostasis. A combination of these processes acts to reduce metal bioavailability, reduce injury to cells and leads to microbial resistance as well as bioremediation potential (Pham *et al.*, 2022). These mechanisms, along with an overview of their underlying molecular interactions and representative bacterial systems, are described in the following sections and Fig. 2.2.

2.5.1   Biosorption

Biosorption is a passive physicochemical process by which living or dead microbial biomass binds and removes HMs from aqueous environments without requiring metabolic energy. Absorption and adsorption are the two different processes used to accumulate metal ions by microbes. Adsorption is a surface phenomenon, whereas absorption involves the whole volume of material. HMs can be easily adsorbed due to the cell wall and mucus layers, and then once through the cell surface they can be absorbed into the cell. Functional

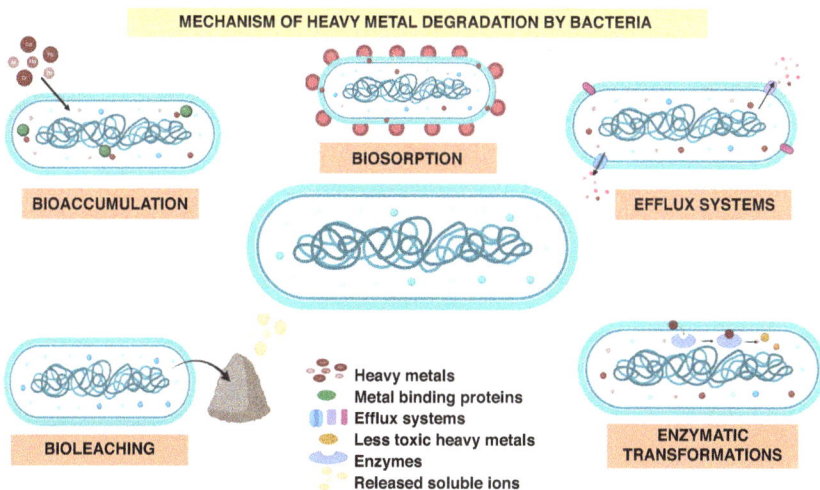

Fig. 2.2. Different mechanisms used by bacteria for degradation of heavy metals.

groups like oxygen, nitrogen, phosphorus and sulfur on the surface of the cell can make complexes with metal ions (Priya *et al.*, 2022). Additionally, HM surfaces carry a cationic group, carboxyl anionic groups and phosphoric acid anions which allow the metal ions to move through the cell membrane. Many bacteria can adsorb HMs onto their cell surfaces or into their biofilms, reducing their bioavailability and toxicity. Biofilms and bacterial extracellular polymeric substances (EPS) make effective barriers and bind HM ions such as Cu, Pb, Cr and Zn (Jin *et al.*, 2018).

In bacteria, biosorption is mainly mediated by the cell wall, which provides numerous functional groups that interact with metal ions. Bacterial cell walls are rich in peptidoglycan, teichoic acids (in Gram-positive bacteria) and lipopolysaccharides (in Gram-negative bacteria), all of which contain functional groups such as carboxyl, phosphate, hydroxyl, amino and sulfhydryl groups. These groups act as binding sites for metal ions through mechanisms like ion exchange, electrostatic attraction, complexation and micro-precipitation. The process occurs rapidly and is not dependent on active transport or energy expenditure, distinguishing it from bioaccumulation (Pham *et al.*, 2022). The ion exchange method, which involves the binding of cell surfaces to HMs, is also a widely used method. Adsorption does not depend upon energy metabolism, but absorption is dependent on energy metabolism (Danis *et al.*, 2008). Adsorption is a quick phenomenon, for example *Bacillus* sp. can adsorb 60% of copper ions in 1 min and reach equilibria in 10 min (He and Tebo, 1998), but absorption is a time consuming and inefficient one. *Citrobacter* sp. JH 11–2 can efficiently remove about 47.7% of cadmium while *Bacillus catenulatus* JB-022 was found to remove about 66% of cadmium (Kim *et al.*, 2015).

Ion exchange is a primary mechanism of bacterial biosorption, wherein cations present in the bacterial cell wall (such as Ca^{2+}, Na^+, K^+ and Mg^{2+}) are replaced by HM ions from the solution. For example, in Gram-negative bacteria like *Pseudomonas aeruginosa*, negatively charged groups on lipopolysaccharides and EPS promote the binding of positively charged metal ions such as Pb^{2+}, Cd^{2+} and Cu^{2+} (Veglio and Beolchini, 1997; Guo *et al.*, 2020).

Various functional groups, including carboxyl and amino groups, form stable complexes on the bacterial cell surface. For example cadmium and copper exhibit a strong binding affinity to amino or phosphate groups in the peptidoglycan layer of *B. subtilis*. The contributions of bacterial EPS play a role in biosorption as they are composed of uronic acids, proteins and polysaccharides applied to the binding of metal ions (Sheng *et al.*, 2008; Hlihor and Gavrilescu, 2009).

In addition to surface adsorption, bacteria can still facilitate localized precipitation of metals on their cell walls. Specifically, when microbial metabolites modify the local pH or become oxidized or reduced, metal salts become nucleated when they are insoluble. For example, *B. cereus* and *P. putida* have both been shown to induce the precipitation of lead and uranium, respectively, through surface-bound phosphate groups and carbonate, leading to a more permanent immobilization mechanism than simple ion exchange (which is reversible) (Park *et al.*, 2011; Ahemad, 2014).

There is a great deal of structural heterogeneity in the cell walls of bacteria, and this has implications for biosorption efficiency. When compared with Gram-negative bacteria, Gram-positive bacteria tend to have a better capacity to bind metals because of their thicker peptidoglycan and teichoic acid layer. However, Gram-negative bacteria secrete more EPS, which may improve the biosorption potential, because EPS give the bacteria additional sites to bind metals. EPS make this easier because they can not only bind metals directly due to the functional groups, but they can also act as a protective barrier that can decrease the toxicity of metals to the bacteria (Pagnanelli *et al.*, 2000; Guibaud *et al.*, 2009). EPS shield the microbes from toxic HMs by limiting entry into the cell. EPS also have anionic and cationic functional groups that accumulate HM ions, including cadmium, copper, cobalt and mercury ions (Sheng *et al.*, 2013; Fang *et al.*, 2017).

2.5.2   Bioaccumulation

Some bacteria actively transport HMs into their cells, then sequester them using metal-binding proteins or peptides like metallothioneins, glutathione or metallochaperones. For example, *Synechococcus* sp. use the *smtA* gene for Cd^{2+} and Zn^{2+} binding. Bioaccumulation in bacteria is the active uptake and intracellular sequestration of dissolved metal ions from the environment (distinct from biosorption, which is passive binding to cell surfaces or dead biomass). In bioaccumulation, the living cell transports metal ions across the membrane and stores/detoxifies them inside the cytoplasm or in intracellular compartments (Pande *et al.*, 2022).

The mechanism of bacterial bioaccumulation involves the active, metabolism-dependent uptake of HM ions from the environment into the intracellular space, where they are detoxified and stored. Bioaccumulation occurs when the absorption rate of contaminants exceeds the rate of loss. Metal ions first interact with the cell surface and are transported across the plasma membrane via specific transporter proteins such as P-type ATPases, NRAMPs (natural resistance associated macrophage proteins) and RND family transporters. Once inside the cytoplasm, metals are sequestered by low-molecular-weight ligands, metallothioneins or phytochelatin-like peptides, which reduce their free ionic toxicity. In some cases, bacteria compartmentalize metals into vacuole-like structures or precipitate them as insoluble granules through biomineralization, while enzymatic transformations such as the reduction of Cr(VI) to Cr(III) further decrease toxicity and mobility. Together, these coordinated processes allow bacteria to survive in metal-stressed environments while effectively removing metals from contaminated water and soils (Mishra and Malik, 2013; Saier, 2016). Following adsorption, bacteria can change HMs into another ionic state in the bacterial cell to decrease their toxicity.

2.5.3 Efflux systems

Bacteria may also use efflux pumps to remove HM ions from the cytoplasm and decrease their concentration inside the cell, thus decreasing toxicity. Bacteria have developed multiple defense mechanisms against the toxic effects of HMs, and efflux is one of the least variable and most investigated mechanisms. Efflux can be defined as the active transport of HM ions from the cytoplasm (or periplasm) to the environment, which prevents the metals from being accumulated intracellularly. These mechanisms involve specific membrane-associated transport mechanisms that are either ATP-dependent or use electrochemical gradients. Efflux mechanisms are mainly responsible for the survival of certain microbes under metal stress and are considered to be a major source of resistance to HMs in bacteria (Nies, 2003).

Bacteria use a range of efflux systems for the removal of HMs (Table 2.1). The five principal groups of efflux systems are ATP-binding cassette (ABC), resistance–nodulation–cell division (RND), major facilitator superfamily (MFS), small multidrug resistance (SMR) and multidrug and toxic compound extrusion (MATE). One of the main types of efflux systems, P-type ATPases, are energy-dependent pumps that hydrolyze ATP to transport metal ions actively into and out of the cytoplasmic membrane. CadA of *Staphylococcus aureus* is a well-characterized cadmium efflux ATPase that exports Cd^{2+}, Zn^{2+} and Pb^{2+} ions across the cytoplasm. Another well-characterized ATPase is CopA, a completely functional copper-exporting ATPase in *Enterococcus hirae*. These systems can provide the initial response of rapid detoxification by lowering HM concentrations in the cytoplasm, which may otherwise complex with and bind to essential biomolecules (Silver and Phung, 2005).

An additional thoroughly researched group of transporters are the RND efflux systems, which are relatively prevalent in Gram-negative bacteria. These transport systems consist of tripartite assemblies that utilize the inner membrane, periplasm and outer membrane to extrude HMs outside of the cell. An example of the CzcCBA system from *Ralstonia metallidurans*, which confers resistance to cadmium, zinc and cobalt, is indicative of a well-characterized system. This system uses three proteins acting in concert with each other for outer membrane export; namely CzcA as the inner membrane transporter, CzcB as the membrane fusion protein and CzcC as the outer membrane channel. RND efflux systems are excellent in environments with high metal concentrations since their broad specificity allows them to bypass the cytoplasmic accumulation of HMs directly (Nies, 1995; Legatzki *et al.*, 2003).

The CDF family is another class of efflux proteins, albeit not ATPases because they are proton–metal antiporters, that expend the proton motive force to expel divalent cations (e.g. Zn^{2+}, Cd^{2+} and Co^{2+}). Well-established CDF examples include the CzcD transporter of *R metallidurans*, and YiiP of *Escherichia coli*. These proteins are also often found to extrude metals into the periplasm (to the cell membrane) and to lessen cytoplasmic toxicity; they also often help RND efflux pumps in the conjugated expulsion of metals (Anton *et al.*, 1999). Some

Table 2.1. Different efflux systems operating in bacteria.

Efflux family/system	Representative pump(s)	Microorganisms	Heavy metals	Key findings	Reference
Resistance–nodulation–division (RND) efflux systems	CzcCBA (divalent cation efflux pump), CusCBA	Gram-negative bacteria	Zn^{2+}, Cd^{2+}, Co^{2+}, Cu^+, Ag^+	Driven by proton motive force Prominent in Gram-negative bacteria Examples include CzcCBA (exports Zn, Cd, Co) and CusCBA (exports Cu, Ag)	Nies, 1995; Chatterjee et al., 2024
P-type ATPases (CPx type)	ZntA, CopA, CadA		Zn^{2+}, Cd^{2+}, Pb^{2+}, Cu	Utilize ATP hydrolysis to actively export heavy metals such as Zn, Cd, Pb and Cu out of the bacterial cell	Balta et al., 2025
Major facilitator superfamily (MFS)		Largest groups of secondary transporters found in all domains of life	Sugars, drugs and toxic heavy metals	Use proton gradients and are involved in the extrusion of various toxic metals and drugs Proteins use the proton motive force (H^+ gradient) or other ion gradients across membranes to mediate the efflux or uptake of small molecules, including small solutes and ions	Gaurav et al., 2023

Continued

Table 2.1. Continued

Efflux family/system	Representative pump(s)	Microorganisms	Heavy metals	Key findings	Reference
Cation diffusion facilitators (CDF); 'ZnT' like	YiiP, ZitB, CzcD	Gram-positive and Gram-negative bacteria	Mediate Zn^{2+} efflux, though some variants are also capable of handling Co^{2+} and Cd^{2+}		Hussein et al., 2023; Bui and Inaba, 2024
Ars systems	Ars operon including acr3 or arsB genes encoding As^{3+} efflux proteins	Gram-positive and Gram-negative bacteria	As(III), Sb(III)	Arsenate [As(V)] is first reduced to arsenite [As(III)] by ArsC, after which efflux transporters actively export As(III) to mitigate toxicity	Yang et al., 2012; Salam et al., 2020
ABC-type efflux	Cad (CadCD/CadDX)	Gram-positive and Gram-negative bacteria	Cd^{2+}	Multi-omics approaches identified Cad-type ABC transporters as major Cd efflux systems in bacterial isolates	Jebril et al., 2022

bacteria also have chromate-specific export systems that are essential in detoxifying the very toxic Cr(VI). The chrA transporter, which has been characterized in several species such as *P. aerugi*nosa and *Cupriavidus metallidurans*, exports chromate from the cytoplasm to the outside. When the export of Cr is coupled with the enzymatic reaction that reduces Cr(VI) to the less toxic Cr(III), the exposure to Cr is further detoxified (Juhnke *et al.*, 2002). Furthermore, members of the MFS contribute to metal resistance. For example, CadB in *S. aureus* works together with CadA ATPase to provide cadmium resistance through the regulation of cadmium uptake versus efflux (Nucifora *et al.*, 1989).

These efflux systems hardly ever work independently as they function cooperatively. An example is *R. metallidurans* in which the RND system CzcCBA operates with the CDF transporter CzcD resulting in effective Zn^{2+}/Cd^{2+} removal from cytoplasmic and periplasmic pools. Efflux pump expression is tightly controlled by metal-sensing transcriptional regulators, such as ArsR, MerR or CadC, that activate or repress efflux genes depending on intracellular metal concentrations (Silver and Phung, 2005).

Efflux mechanisms are vital for the survival of bacteria in environments with heavy contamination including mining soils, wastewaters and industrial effluents. Their efficiency makes them attractive targets for bioremediation, and genetically engineered strains with overexpressed efflux pumps are being investigated for wastewater treatment and detoxification of polluted sites. Furthermore, the fact that efflux genes are often plasmid-borne enhances their horizontal transfer across microbial communities, enabling rapid adaptation to HM stress (Nies, 2003; Seiler and Berendonk, 2012).

2.5.4 Enzymatic transformation

Certain bacteria can enzymatically convert HMs into less toxic forms that are less soluble, volatile or more easily immobilized using specific enzymes. Different bacteria use different enzymes to decrease the availability of different HMs. Some of the common enzymes reported in different studies are given in Table 2.2.

2.5.5 Bioleaching

Processes such as biological dissolution or complexation processes and biooxidation are used to mobilize HMs. The dissolution of HMs can occur due to microbial secretions. Microbes can promote the leaching of cadmium by using nutrients and energy.

In precipitation and biomineralization some bacterial species produce metabolites like hydrogen sulfide, which react with HMs to form insoluble metal sulfides (e.g. CdS), precipitating them out of solution and effectively immobilizing them.

Table 2.2. Enzymes involved in heavy metal removal.

Heavy metals	Enzymes	Mechanism	Reference
Mercury	Mercuric reductase Organomercurial lyase	Volatilization of Hg, reducing Hg^{+2} to Hg^0 and releasing it into the environment	Priyadarshanee et al., 2022
Arsenic	Arsenate reductase	Catalyzes the conversion of AsV to AsIII. After the conversion, the enzyme facilitates the efflux of AsIII from the cell	Mohsin et al., 2023
Chromium	Chromate reductases (ChrA and YieF) NADH-dependent nitroreductase Iron reductase Quinone reductases Hydrogenases NADH/NADPH-dependent flavin reductases NADPH-dependent reductases	Reduction of Cr(VI) to Cr(III)	Rahman and Thomas, 2021
Manganese	Multi-copper oxidases and Mn oxidases	Immobilize manganese as insoluble minerals	Naziębłо and Dobrzyński, 2025

In mineralization, bacteria remove HMs by producing anions or reductants that convert soluble metal ions into insoluble mineral phases: carbonate (microbially induced carbonate precipitation (MICP)), phosphate (microbially induced phosphate precipitation (MIPP)), sulfide (microbially induced sulfide precipitation (MISP)) or reduced oxides/sulfides via enzymatic redox.

Cells/EPS provide nucleation templates; environmental parameters (pH, redox, anion availability) determine which mineral forms and how stable the immobilization will be.

These biomineralization routes have been shown in the last few years to immobilize critical metals including Pb, Cd, Zn, uranium and others with high efficiency, and are being translated into bioreactor, soil stabilization and *in situ* remediation applications (Mallick *et al.*, 2025). Chen *et al.* (2025) investigated the role of sulfate-reducing bacteria *Desulfovibrio desulfuricans* and *Desulfobulbus propionicus* in Cd^{2+} and Pb^{2+} biomineralization. *D. propionicus* showed higher Cd^{2+} immobilization (up to 98.97%) compared with *D. desulfuricans* (75.62%), while both achieved *c*.80% Pb^{2+} removal at concentrations <50 mg/L. Mechanistic analyses (SEM-EDS (scanning electron microscopy–energy-dispersive X-ray spectroscopy), TEM (transmission electron microscopy), XRD (X-ray diffraction), Raman and XPS (X-ray photoelectron spectroscopy)) revealed Cd^{2+} precipitation mainly as CdS, while Pb^{2+} formed phosphate, oxide and sulfide precipitates. These findings highlight sulphate-reducing bacteria (SRB) mediated sulfate reduction as an efficient, low-cost bioremediation strategy for HM-contaminated mine drainage.

2.6 Role of Different Bacteria in Bioremediation

Bacteria belonging to different genera are actively involved in decreasing the availability of metals. Aerobic bacteria decontaminate sites by using pollutants like HMs, pesticides and hydrocarbon compounds as their carbon and energy sources (e.g. *Mycobacterium*, *Pseudomonas putida*, *P. diminuts*, *Sphingomonas* and *Rhodococcus*). Although anaerobic bacteria do not occur as frequently as aerobic bacteria, their *in situ* approach primarily focuses on anaerobic organisms (Elgh Dalgren *et al.*, 2009). These bacteria develop different mechanisms to adapt to HM exposure, which provides a greater understanding of microbial resilience and a basis for studying the mechanisms of HM tolerance in contaminated soil (Du *et al.*, 2025).

2.6.1 *Bacillus* species

Due to their diverse metabolic pathways, resistance and endospores, *Bacillus* sp. are some of the best microorganisms for the bioremediation of HM-contaminated sites. *Bacillus* sp. can use different metabolic pathways to bioremediate metals. Several studies have demonstrated the ability of *Bacillus* sp. to bioremediate HMs including Pb, Cd, Zn, Hg, Cr, As and Ni. *Bacillus* sp. – particularly *B. subtilis*, *B. cereus* and *B. thuringiensis* – are widely recognized promising agents for the removal of HM pollution in diverse environments. *Bacillus* sp. can remove toxic HMs by biosorption where HM ions are attached to their cell surfaces via functional groups, hence reducing metal concentrations in soils and water. Other species can reduce the bioavailability of HMs in the environment by sequestering them within their cells using phenomenon known as bioaccumulation. *Bacillus* sp. prevent the dispersion of HMs by precipitation by converting dissolved metals into insoluble forms through various chemical transformations. They can produce siderophores and enzymes that solubilize and chelate metals, aiding in the detoxification and sequestration process (Wróbel *et al.*, 2023). *B. cereus* strains have achieved Cd removal rates of up to 81%, Pb up to 40% and Zn up to 38% under laboratory conditions; the production of iron carriers might be the reason for the Cd tolerance of different strains (Liang *et al.*, 2025).

    Aluminium (Al) toxicity in acidic soils severely restricts crop growth by inhibiting root development and nutrient uptake. An acid-tolerant *B. subtilis* isolate (MBB3B9) was found to mitigate Al stress in rice by producing siderophores, organic acids, phytohormones and solubilizing essential nutrients.

Genome analysis revealed pathways for organic acid production, indole-3-acetic acid (IAA) and cytokinin biosynthesis, and nutrient metabolism as key mechanisms. Pot experiments under acidic conditions showed that MBB3B9 treatment significantly improved rice growth, yield and root biomass compared with untreated controls (Hazarika *et al.*, 2023).

2.6.2 *Pseudomonas* species

Pseudomonas sp. play a significant and versatile role in the bioremediation of HM-contaminated environments. These bacteria are well studied for their natural metal resistance, capacity to remove and detoxify metals and their adaptability across environmental conditions. Due to their metabolic diversity, *Pseudomonas* sp. have been reported to remove HMs such as Pb, Cd, Ni, Cr and Hg by biosorption in which metal ions can attach and immobilize to the bacterial cell surfaces, leveraging cell wall components and exopolysaccharides for removal. Certain strains can internalize metals, sequestering them within cellular compartments, using the phenomenon of bioaccumulation. Some *Pseudomonas* sp. uses enzymatic transformation using specific enzymes (e.g. chromate reductase in *P. putida*, mer-operon for mercury reduction) that chemically transform metals into less toxic forms, notably reducing Cr(VI) to Cr(III) and detoxifying mercury. The excreted biofilm-forming polysaccharides aid in binding and aggregating metal ions, enhancing removal efficacy. Mercury-resistant strains of *P. aeruginosa* absorb mercury ions selectively with a maximum uptake capacity of about 180 mg/g (Yin *et al.*, 2016).

Pseudomonas putida SP1 is resistant to mercury and can absorb 100% of mercury in the marine environment and then reduce toxic Hg(II) to Hg^0 by enzyme reductase, which envisages the possibility of bioremediation of mercury intoxication (Zhang *et al.*, 2012). Robas Mora *et al.* (2022) showed that a *Pseudomonas* sp. strain SAICEUPSM^T conferred oxidative-stress protection and phytoprotection to plants under mercury stress, highlighting dual benefits for remediation and agriculture. Mixed bacterial cultures of *Cupriavidus metallidurans* LBJ and *P. stutzeri* LBR, isolated from Pb-contaminated soils, achieved substantial Pb reduction under non-sterile conditions (Ridene *et al.*, 2023).

Vélez *et al.* (2021) screened *Pseudomonas* sp., including *P. aeruginosa* and *P. nitroreducens*, documenting high Pb biosorption capacities and processing parameters relevant to wastewater treatment. Mei *et al.* (2024) characterized a highly Cd(II)-resistant *P. aeruginosa* that upregulated pyoverdine under Cd stress, linking siderophore production to resistance and uptake dynamics. Chatterjee *et al.* (2024) used comparative genomics to map multimodal Cd resistance and regulatory evolution across *Pseudomonas* sp., explaining the frequent dominance of *P. aeruginosa* in contaminated environments. Yang *et al.* (2022) described *P. stutzeri* YC-34 as an auto-aggregating aerobic denitrifier capable of simultaneous Cr(VI) biosorption and reduction alongside nitrogen removal. John and Rajan (2022) optimized bioreactor conditions for Cr(VI) reduction by *P. putida* strain APRRJVITS11 using tannery effluent as a test matrix. Benites-Alfaro *et al.* (2025) further evaluated biological Cr(VI) reduction by *P. putida*, reporting robust detox over multi-week operations. Ramli *et al.* (2023) reviewed metabolic pathways for bacterial Cr(VI) reduction with specific links to *Pseudomonas* enzymes and electron-transfer routes. Xue *et al.* (2024) demonstrated arsenic removal from mining wastewater using a controllable genetically modified organism (GMO) *Pseudomonas* integrated with biochar, offering a process-level blueprint for scalable As remediation. Elsayed *et al.* (2022) modeled Co(II) biosorption by *P. alcaliphila* NEWG-2, comparing artificial neural network (ANN) based predictions with design-of-study optimization to map key process factors. Qurbani *et al.* (2025) showed that a microbial consortium including *Pseudomonas* members enhanced tolerance and removal of Cu, Zn and Ni, underscoring the advantages of community-level approaches.

2.6.3 *Arthrobacter* species

Batch experiments using dead and living *Arthrobacter viscosus* biomass demonstrated complete reduction of Cr(VI) to Cr(III) at acidic pH (1–2) for concentrations below $100\,mg/L$. Cr removal followed the Langmuir isotherm and kinetic models, with biosorbent dosages of 5–8 g/L. In an open system, biofilm-supported biomass achieved a Cr uptake of $20.37\,mg/g$, indicating strong scalability. These findings confirm *A. viscosus* as an efficient and sustainable option for treating chromium-contaminated wastewater (Hlihor *et al.*, 2017).

Eco-friendly management of e-waste through biohydrometallurgy was explored using *Arthrobacter* sp. EIKU3, a metal-resistant bacterium isolated from metalliferous soil. The strain showed strong potential for Cr reduction (>88% at pH 7.0) and Cu sequestration, removing up to 82% Cu from waste-printed circuit board (WPCB) leachate. Biomass grown in both minimal media and municipal wastewater efficiently removed Cu, with most of the absorbed metal localized in the cytoplasm. These results highlight *Arthrobacter* sp. EIKU3 as a promising candidate for simultaneous pollution mitigation and metal recovery from e-waste (Chakraborty *et al.* (2025).

Chromium contamination from industrial waste poses major environmental risks, and microbially induced carbonate precipitation (MICP) offers a sustainable remediation strategy. In one study, *A. creatinolyticus* reduced toxic Cr(VI) to Cr(III) through EPS production and facilitated co-precipitation with $CaCO_3$, immobilizing over 82% of Cr. MICP treatment in sand and soil significantly decreased the exchangeable fraction of Cr while increasing its carbonate-bound form. Structural analyses confirmed Cr(III)-mediated polymorph selection of vaterite, demonstrating the potential of *A. creatinolyticus* for stabilizing Cr in contaminated soils and preventing groundwater pollution (Sujiritha *et al.*, 2024).

Pathak *et al.* (2020) examined the genomic characterization of a mercury-resistant *Arthrobacter* sp. H-02–3 isolated from Savannah River wetland soil. Industrial activities at the Savannah River site, South Carolina, have led to Hg contamination in the H-02 wetland system. From these soils, a highly Hg-resistant bacterium, *Arthrobacter* sp. H-02–3, was isolated and sequenced, revealing genes linked to metal resistance, including *merA* and *merB* that enable MeHg demethylation and the reduction of Hg^{2+} to volatile Hg^0. Comparative genomics showed 1100 unique genes, many associated with metal resistance, alongside co-occurring antibiotic resistance genes. These findings highlight the bacterium's role in Hg detoxification and raise concerns about artificial and metal resistance gene (ARG–MRG) co-selection in contaminated environments.

2.6.4 *Rhodobacter* species

A study by Al-Ansari (2022) evaluated the eco-friendly remediation of As using As-resistant *Rhodobacter* sp. The strain showed tolerance up to 500 mg/l As, with optimal growth under light blue illumination and maximum As removal (87.5%) at pH 7.0 with glucose and citrate supplementation. Gene expression analysis revealed upregulation of *arsC* and *aio* (involved in As transformation) between 24 and 72 h, followed by downregulation at 96 h. These findings highlight *Rhodobacter* sp. as a promising candidate for As bioremediation (Al-Ansari, 2022).

Liu *et al.* (2023) suggested that *R. sphaeroides* can be used as a model to study the toxicity of to HM ions and other contaminants. The organism showed increased growth inhibition with chain length, accompanied by membrane damage, altered carotenoid absorption and blue shift in the B850 band of light-harvesting complex 2. This study examined HMs and polychlorinated biphenyls (PCBs) in contaminated soil from Taranto, Italy, and tested plant-assisted bioremediation using a model plant *Arabidopsis thaliana*. While pollutants negatively affected plant growth, supplementation with *R. sphaeroides* improved tolerance by stimulating HM-associated proteins (notably AtHMP23) and reducing chlorophyll loss through downregulation of the magnesium-dechelatase gene. The results demonstrate the potential of *R. sphaeroides* to enhance plant resilience and support soil restoration in multi-contaminated environments (Labarile *et al.*, 2025).

In a study conducted in aquaculture (Zhou *et al.*, 2025), it was reported that *R. sphaeroides* SC01 – a purple non-sulfur photosynthetic bacteria, a probiotic candidate with strong Pb tolerance – colonizes the intestine of the common carp under Pb stress, improving gut microbiota diversity and beneficial metabolites. Supplementation with SC01 reduced Pb accumulation in tissues, alleviated intestinal and liver damage, enhanced antioxidant and immune responses, and improved growth performance, indicating its promise as a dietary supplement for mitigating Pb toxicity in aquaculture and pointing to applications in reducing HM bioaccumulation in food chains (Zhou *et al.*, 2025).

Rhodobacter capsulatus can adsorb Zn(II) with a maximum uptake capacity of nearly 164 mg/g, following the Langmuir and Redlich–Peterson isotherms (Magnin *et al.*, 2014).

Biofilms of *Staphylococcus epidermis* eliminate Cr(VI) with high removal efficacy, following the Quindrich isotherm (Quiton *et al.*, 2018). A study conducted by Bhattacharya and Gupta (2013) reported a newly isolated strain (B9) of *Acinetobacter* sp. with the potential for detoxifying Cr released from the metal furnishing industry. The results further confirmed that the isolated strain was capable of tolerating up to 350 mg/L of Cr(VI) and also showed a level of tolerance to Ni(II), Zn(II), Pb(II) and Cd(II). Furthermore, it was able to remove up to 67% of Cr(VI) (concentration 7.0 mg/L) within 24 hr (Bhattacharya and Gupta, 2013). In another study, almost 72 acidothermophilic autotrophic microbes were screened for their metal tolerance and biosorption potentiality. The results confirmed that the ATh-14 strain was efficient for solubilization of copper with 85.82% efficiency in the presence of 10^{-3} M multi-metal concentration within 5 days (Umrania, 2006).

2.7 Conclusion

Heavy metal contamination from different sources poses a major environmental risk which is leading to many serious health problems in animals and humans. Bacteria can convert soluble metal ions into stable and insoluble mineral phases using various mechanisms, thereby reducing their mobility and bioavailability. Bioremediation of HMs using bacteria is evolving to overcome limitations such as efficiency and scale-up issues. Recently, advancements in the fields of molecular biology, synthetic and nanotechnology have dramatically improved the performance and utility of bacterial bioremediation in the decantation of HMs. Simultaneously, synthetic biology technologies have outlined scalable design approaches to design and build sensors and detoxifiers of bacteria capable of sensing and eliminating HMs in polluted media. The application of nanotechnology, omics-based tools and process engineering involving bacterial systems continue to enhance the recovery of heavy metals and to aid biomass separation following remediation (Zhang *et al.*, 2025).

References

Ahemad, M. (2014) Bacterial mechanisms for cr(VI) resistance and reduction: An overview and recent advances. *Folia Microbiologica* 59(4), 321–332. DOI: 10.1007/s12223-014-0304-8.

Al-Ansari, M.M. (2022) Influence of blue light on effective removal of arsenic by photosynthetic bacterium *Rhodobacter* sp. BT18. *Chemosphere* 292, 133399. DOI: 10.1016/j.chemosphere.2021.133399.

Anton, A., Große, C., Reißmann, J., Pribyl, T. and Nies, D.H. (1999) CzcD is a heavy metal ion transporter involved in regulation of heavy metal resistance in *Ralstonia* sp. strain CH34 . *Journal of Bacteriology* 181(22), 6876–6881. DOI: 10.1128/ JB.181.22.6876-6881.1999.

Atlas, R.M. and Philp, J. (eds) (2005) *Bioremediation. Applied Microbial Solutions for Real-World Environmental Cleanup.* American Society of Microbiology. DOI: 10.1128/9781555817596.

Aziz, K.H.H., Mustafa, F.S., Omer, K.M., Hama, S., Hamarawf, R.F. *et al.* (2023) Heavy metal pollution in the aquatic environment: Efficient and low-cost removal approaches to eliminate their toxicity: A review. *RSC Advances* 13(26), 17595–17610. DOI: 10.1039/D3RA00723E.

Balali-Mood, M., Naseri, K., Tahergorabi, Z., Khazdair, M.R. and Sadeghi, M. (2021) Toxic mechanisms of five heavy metals: Mercury, lead, chromium, cadmium, and arsenic. *Frontiers in Pharmacology* 12, 643972. DOI: 10.3389/ fphar.2021.643972.

Balta, I., Lemon, J., Gadaj, A., Cretescu, I., Stef, D. *et al.* (2025) The interplay between antimicrobial resistance, heavy metal pollution, and the role of microplastics. *Frontiers in Microbiology* 16, 1550587. DOI: 10.3389/fmicb.2025.1550587.

Benites-Alfaro, E., Aguinaga, D.L., Carranza, C.C., Nakayo, J.J., Suca-Apaza, G.R. *et al.* (2025) Biological method with *Pseudomonas putida* for chromium VI reduction in chromium plating process wastewater. *Chemical Engineering Transactions* 117, 427–432.

Bhattacharya, A. and Gupta, A. (2013) Evaluation of *Acinetobacter* sp. B9 for Cr (VI) resistance and detoxification with potential application in bioremediation of heavy-metals-rich industrial wastewater. *Environmental Science and Pollution Research* 20(9), 6628–6637. DOI: 10.1007/s11356-013-1728-4.

Briffa, J., Sinagra, E. and Blundell, R. (2020) Heavy metal pollution in the environment and their toxicological effects on humans. *Heliyon* 6(9), e04691. DOI: 10.1016/j. heliyon.2020.e04691.

Bui, H.B. and Inaba, K. (2024) Structures, mechanisms, and physiological functions of zinc transporters in different biological kingdoms. *International Journal of Molecular Sciences* 25(5), 3045. DOI: 10.3390/ijms25053045.

Chakraborty, A., Bhakat, K., Islam, E. and Murmu, R. (2025) *Arthrobacter* sp. mediated chromium remediation and copper accumulation from leached liquor for e-waste management. *The Microbe* 6, 100277. DOI: 10.1016/j.microb.2025.100277.

Chatterjee, S., Barman, P., Barman, C., Majumdar, S. and Chakraborty, R. (2024) Multimodal cadmium resistance and its regulatory networking in *Pseudomonas aeruginosa* strain CD3. *Scientific Reports* 14(1), 31689. DOI: 10.1038/ s41598-024-80754-y.

Chen, Q., Min, Q., Wu, H., Zhang, L. and Si, Y. (2025) Biomineralization of Cd^{2+} and Pb^{2+} by sulfate-reducing bacteria *Desulfovibrio desulfuricans* and *Desulfobulbus propionicus*. *Frontiers in Environmental Science* 13, 1591564. DOI: 10.3389/ fenvs.2025.1591564.

Dagdag, O., Quadri, T.W., Haldhar, R., Kim, S.C., Daoudi, W. *et al.* (2023) An overview of heavy metal pollution and control. In: *Heavy Metals in the Environment: Management Strategies for Global Pollution.* pp. 3–24. DOI: 10.1021/bk-2023-1456.ch001.

Danis, U., Nuhoglu, A. and Demirbas, A. (2008) Ferrous ion-oxidizing in *Thiobacillus ferrooxidans* batch cultures: Influence of pH, temperature and initial concentration of Fe^{2+}. *Fresenius Environment Bulletin* 17, 371–377.

Du, J., Yuan, Y., Li, J., Zhang, S. and Ren, Y. (2025) Preliminary study on mercury pollution affecting soil bacteria near a mercury mining area. *Frontiers in Microbiology* 16, 1539059. DOI: 10.3389/fmicb.2025.1539059.

Elgh Dalgren, K., Waara, S., Düker, A., von Kronhelm, T. and van Hees, P.A.W. (2009) Anaerobic bioremediation of a soil with mixed contaminants: Explosives degradation and influence on heavy metal distribution, monitored as changes in concentration and toxicity. *Water, Air, and Soil Pollution* 202(1–4), 301–313. DOI: 10.1007/s11270-009-9977-z.

Elsayed, A., Moussa, Z., Alrdahe, S.S., Alharbi, M.M., Ghoniem, A.A. *et al.* (2022) Optimization of heavy metals biosorption via artificial neural network: A case study of Cobalt (II) sorption by *Pseudomonas alcaliphila* NEWG-2. *Frontiers in Microbiology* 13, 893603. DOI: 10.3389/fmicb.2022.893603.

Evans, G.M. and Furlong, J.C. (2003) *Environmental Biotechnology Theory and Application.* UK: Wiley Chichester.

Fang, X., Li, J., Li, X., Pan, S., Zhang, X. *et al.* (2017) Internal pore decoration with polydopamine nanoparticle on polymeric ultrafiltration membrane for enhanced heavy metal removal. *Chemical Engineering Journal* 314, 38–49. DOI: 10.1016/j.cej.2016.12.125.

Gaurav, A., Bakht, P., Saini, M., Pandey, S. and Pathania, R. (2023) Role of bacterial efflux pumps in antibiotic resistance, virulence, and strategies to discover novel efflux pump inhibitors. *Microbiology* 169(5), 001333.

Gol-Soltani, M., Ghasemi-Fasaei, R., Ronaghi, A., Zarei, M., Zeinali, S. *et al.* (2024) Efficient immobilization of heavy metals using newly synthesized magnetic nanoparticles and some bacteria in a multi-metal contaminated soil. *Environmental Science and Pollution Research* 31(27), 39602–39624.

Guibaud, G., van Hullebusch, E., Bordas, F., d'Abzac, P. and Joussein, E. (2009) Sorption of Cd(II) and Pb(II) by exopolymeric substances (EPS) extracted from activated sludges and pure bacterial strains: Modeling of the metal/ligand ratio effect and role of the mineral fraction. *Bioresource Technology* 100(12), 2959–2968. DOI: 10.1016/j.biortech.2009.01.040.

Guo, B., Hong, C., Tong, W., Xu, M., Huang, C. *et al.* (2020) Health risk assessment of heavy metal pollution in a soil-rice system: A case study in the Jin-Qu Basin of China. *Scientific Reports* 10(1), 11490. DOI: 10.1038/s41598-020-68295-6.

Hazarika, D.J., Bora, S.S., Naorem, R.S., Sharma, D., Boro, R.C. *et al.* (2023) Genomic insights into *Bacillus subtilis* MBB3B9 mediated aluminium stress mitigation for enhanced rice growth. *Scientific Reports* 13(1), 16467. DOI: 10.1038/s41598-023-42804-9.

He, L.M. and Tebo, B.M. (1998) Surface charge properties of and Cu(II) adsorption by spores of the marine *Bacillus* sp. Strain SG-1. *Applied and Environmental Microbiology* 64(3), 1123–1129. DOI: 10.1128/AEM.64.3.1123-1129.1998.

Hlihor, R.M. and Gavrilescu, M. (2009) Removal of some environmentally relevant heavy metals using low-cost natural sorbents. *Environmental Engineering and Management Journal* 8(2), 353–372. DOI: 10.30638/eemj.2009.051.

Hlihor, R.M., Figueiredo, H., Tavares, T. and Gavrilescu, M. (2017) Biosorption potential of dead and living *Arthrobacter viscosus* biomass in the removal of Cr(VI): Batch and column studies. *Process Safety and Environmental Protection* 108, 44–56. DOI: 10.1016/j.psep.2016.06.016.

Hussein, A., Fan, S., Lopez-Redondo, M., Kenney, I., Zhang, X. *et al.* (2023) Energy coupling and stoichiometry of Zn^{2+}/H^+ antiport by the prokaryotic cation diffusion facilitator YiiP. *eLife* 12, RP87167. DOI: 10.7554/eLife.87167.

Ilyas, T., Shahid, M., Shafi, Z. and Aijaz, S.A. (2025) Molecular mechanisms of methyl jasmonate (MeJAs)-mediated detoxification of heavy metals (HMs) in agricultural crops: An interactive review. *South African Journal of Botany* 177, 139–159. DOI: 10.1016/j.sajb.2024.11.031.

Jaishankar, M., Tseten, T., Anbalagan, N., Mathew, B.B. and Beeregowda, K.N. (2014) Toxicity, mechanism and health effects of some heavy metals. *Interdisciplinary Toxicology* 7(2), 60–72. DOI: 10.2478/intox-2014-0009.

Jan, A.T., Azam, M., Siddiqui, K., Ali, A., Choi, I. *et al.* (2015) Heavy metals and human health: Mechanistic insight into toxicity and counter defense system of antioxidants. *International Journal of Molecular Sciences* 16(12), 29592–29630. DOI: 10.3390/ijms161226183.

Jebril, N., Boden, R. and Braungardt, C. (2022) Cadmium resistant bacteria mediated cadmium removal: A systematic review on resistance, mechanism and bioremediation approaches. In: *IOP Conference Series: Earth and Environmental Science*, Vol. 1002. IOP Publishing, p. 012006. DOI: 10.1088/1755-1315/1002/1/012006.

Jin, Y., Luan, Y., Ning, Y. and Wang, L. (2018) Effects and mechanisms of microbial remediation of heavy metals in soil: A critical review. *Applied Sciences* 8(8), 1336. DOI: 10.3390/app8081336.

John, R. and Rajan, A.P. (2022) Bioreactor level optimization of chromium (VI) reduction through *Pseudomonas putida* APRRJVITS11 and sustainable remediation of pathogenic DNA in water. *Beni-Suef University Journal of Basic and Applied Sciences* 11(1), 13. DOI: 10.1186/s43088-021-00183-y.

Jomova, K., Alomar, S.Y., Nepovimova, E., Kuca, K. and Valko, M. (2025) Heavy metals: Toxicity and human health effects. *Archives of Toxicology* 99(1), 153–209. DOI: 10.1007/s00204-024-03903-2.

Juhnke, S., Peitzsch, N., Hübener, N., Grosse, C. and Nies, D.H. (2002) New genes involved in chromate resistance in *Ralstonia metallidurans* strain CH34. *Archives of Microbiology* 179(1), 15–25. DOI: 10.1007/s00203-002-0492-5.

Karnwal, A., Kumar, G., Din Mahmoud, A.E., Dutta, J., Singh, R. *et al.* (2025) Eco-engineered remediation: Microbial and rhizosphere-based strategies for heavy metal detoxification. *Current Research in Biotechnology* 9, 100297. DOI: 10.1016/j.crbiot.2025.100297.

Kim, J.J., Kim, Y.S. and Kumar, V. (2019) Heavy metal toxicity: An update of chelating therapeutic strategies. *Journal of Trace Elements in Medicine and Biology* 54, 226–231. DOI: 10.1016/j.jtemb.2019.05.003.

Kim, S.Y., Jin, M.R., Chung, C.H., Yun, Y.-S., Jahng, K.Y. *et al.* (2015) Biosorption of cationic basic dye and cadmium by the novel biosorbent *Bacillus catenulatus* JB-022 strain. *Journal of Bioscience and Bioengineering* 119(4), 433–439. DOI: 10.1016/j.jbiosc.2014.09.022.

Labarile, R., Cotugno, P., Ancona, V., Trotta, M. and Veronico, P. (2025) Biostimulation effect of Rhodobacter sphaeroides on *Arabidopsis thaliana* grown in soils contaminated with heavy metals and polychlorinated biphenyls. *Current Plant Biology* 42, 100486. DOI: 10.1016/j.cpb.2025.100486.

Legatzki, A., Grass, G., Anton, A., Rensing, C. and Nies, D.H. (2003) Interplay of the Czc system and two P-type ATPases in conferring metal resistance to *Ralstonia metallidurans*. *Journal of Bacteriology* 185(15), 4354–4361.

Liang, B., Feng, Y., Ji, X., Li, C., Li, Q. *et al.* (2025) Isolation and characterization of cadmium-resistant *Bacillus cereus* strains from Cd-contaminated mining areas for potential bioremediation applications. *Frontiers in Microbiology* 16, 1550830.

Liu, X.-L., Chen, M.-Q., Jiang, Y.-L., Gao, R.-Y., Wang, Z.-J. *et al.* (2023) *Rhodobacter sphaeroides* as a model to study the ecotoxicity of 1-alkyl-3-methylimidazolium bromide. *Frontiers in Molecular Biosciences* 10, 1106832. DOI: 10.3389/fmolb.2023.1106832.

Magnin, J.-P., Gondrexon, N. and Willison, J.C. (2014) Zinc biosorption by the purple non-sulfur bacterium *Rhodobacter capsulatus*. *Canadian Journal of Microbiology* 60(12), 829–837. DOI: 10.1139/cjm-2014-0231.

Mallick, S., Pradhan, T. and Das, S. (2025) Bacterial biomineralization of heavy metals and its influencing factors for metal bioremediation. *Journal of Environmental Management* 373, 123977. DOI: 10.1016/j.jenvman.2024.123977.

Mei, S., Bian, W., Yang, A., Xu, P., Qian, X. *et al.* (2024) The highly effective cadmium-resistant mechanism of *Pseudomonas aeruginosa* and the function of pyoverdine induced by cadmium. *Journal of Hazardous Materials* 469, 133876. DOI: 10.1016/j.jhazmat.2024.133876.

Mishra, A. and Malik, A. (2013) Recent advances in microbial metal bioaccumulation. *Critical Reviews in Environmental Science and Technology* 43(11), 1162–1222. DOI: 10.1080/10934529.2011.627044.

Mohsin, H., Shafique, M., Zaid, M. and Rehman, Y. (2023) Microbial biochemical pathways of arsenic biotransformation and their application for bioremediation. *Folia Microbiologica* 68(4), 507–535. DOI: 10.1007/s12223-023-01068-6.

Naziębło, A. and Dobrzyński, J. (2025) Biotransformation of As, Cr, Hg, and Mn by *Pseudomonadota*: Chances and risks. *Biodegradation* 36(4), 1–28. DOI: 10.1007/s10532-025-10157-x.

Nies, D.H. (1995) The cobalt, zinc, and cadmium efflux system CzcABC from *Alcaligenes eutrophus* functions as a cation-proton antibporter in *Eschericia coli*. *Journal of Bacteriology* 177(10), 2707–2712. DOI: 10.1128/jb.177.10.2707-2712.1995.

Nies, D.H. (2003) Efflux-mediated heavy metal resistance in prokaryotes. *FEMS Microbiology Reviews* 27(2–3), 313–339.

Nnaji, N.D., Onyeaka, H., Miri, T. and Ugwa, C. (2023) Bioaccumulation for heavy metal removal: A review. *SN Applied Sciences* 5(5), 125. DOI: 10.1007/s42452-023-05351-6.

Nucifora, G., Chu, L., Misra, T.K. and Silver, S. (1989) Cadmium resistance from *Staphylococcus aureus* plasmid pI258 cadA gene results from a cadmium-efflux ATPase. *Proceedings of the National Academy of Sciences* 86(10), 3544–3548. DOI: 10.1073/pnas.86.10.3544.

Pagnanelli, F., Petrangeli Papini, M., L.Toro, Trifoni, M., Vegliò, F. *et al.* (2000) Biosorption of metal ions on *Arthrobacter* sp.: Biomass characterization and biosorption modeling. *Environmental Science & Technology* 34(13), 2773–2778. DOI: 10.1021/es991271g.

Pande, V., Pandey, S.C., Sati, D., Bhatt, P. and Samant, M. (2022) Microbial interventions in bioremediation of heavy metal contaminants in agroecosystem. *Frontiers in Microbiology* 13, 824084. DOI: 10.3389/fmicb.2022.824084.

Park, J.H., Lamb, D., Paneerselvam, P., Choppala, G., Bolan, N. *et al.* (2011) Role of organic amendments on enhanced bioremediation of heavy metal(loid) contaminated soils. *Journal of Hazardous Materials* 185(2–3), 549–574. DOI: 10.1016/j.jhazmat.2010.09.082.

Pathak, A., Jaswal, R. and Chauhan, A. (2020) Genomic characterization of a mercury resistant *Arthrobacter* sp. H-02-3 reveals the presence of heavy metal and antibiotic resistance determinants. *Frontiers in Microbiology* 10. DOI: 10.3389/fmicb.2019.03039.

Pham, V.H.T., Kim, J., Chang, S. and Chung, W. (2022) Bacterial biosorbents, an efficient heavy metals green clean-up strategy: Prospects, challenges, and opportunities. *Microorganisms* 10(3), 610. DOI: 10.3390/microorganisms10030610.

Priya, A.K., Gnanasekaran, L., Dutta, K., Rajendran, S., Balakrishnan, D. *et al.* (2022) Biosorption of heavy metals by microorganisms: Evaluation of different underlying mechanisms. *Chemosphere* 307(Pt 4), 135957. DOI: 10.1016/j.chemosphere.2022.135957.

Priyadarshanee, M., Chatterjee, S., Rath, S., Dash, H.R. and Das, S. (2022) Cellular and genetic mechanism of bacterial mercury resistance and their role in biogeochemistry and bioremediation. *Journal of Hazardous Materials* 423(Pt A), 126985. DOI: 10.1016/j.jhazmat.2021.126985.

Quiton, K.G., Doma, B., Jr., Futalan, C.M. and Wan, M.-W. (2018) Removal of chromium (VI) and zinc (II) from aqueous solution using kaolin-supported bacterial biofilms of Gram-negative *E. coli* and Gram-positive *Staphylococcus epidermidis*. *Sustainable Environment Research* 28(5), 206–213. DOI: 10.1016/j.serj.2018.04.002.

Qurbani, K., Wsw, H., Khdhr, R., Hussein, S., Ibrahim, B. *et al.* (2025) Synergistic enhancement of heavy metal tolerance and reduction by indigenous bacterial consortia of *Pseudomonas putida* and *Pasteurella aerogenes*. *Scientific Reports* 15(1), 24663. DOI: 10.1038/s41598-025-99238-8.

Rahman, Z. and Thomas, L. (2021) Chemical-assisted microbially mediated chromium (Cr)(VI) reduction under the influence of various electron donors, redox mediators, and other additives: An outlook on enhanced Cr (VI) removal. *Frontiers in Microbiology* 11, 619766. DOI: 10.3389/fmicb.2020.619766.

Ramli, N.N., Othman, A.R., Kurniawan, S.B., Abdullah, S.R.S. and Hasan, H.A. (2023) Metabolic pathway of Cr(VI) reduction by bacteria: A review. *Microbiological Research* 268, 127288. DOI: 10.1016/j.micres.2022.127288.

Ridene, S., Werfelli, N., Mansouri, A., Landoulsi, A. and Abbes, C. (2023) Bioremediation potential of consortium *Pseudomonas Stutzeri* LBR and *Cupriavidus Metallidurans* LBJ in soil polluted by lead. *PloS ONE* 18(6), e0284120. DOI: 10.1371/journal.pone.0284120.

Robas Mora, M., Fernández Pastrana, V.M., González Reguero, D., Gutiérrez Oliva, L.L., Probanza Lobo, A. *et al.* (2022) Oxidative stress protection and growth promotion activity of *Pseudomonas mercuritolerans* sp. nov., in forage plants under mercury abiotic stress conditions. *Frontiers in Microbiology* 13, 1032901. DOI: 10.3389/fmicb.2022.1032901.

Roy, M., Giri, A.K., Dutta, S. and Mukherjee, P. (2015) Integrated phytobial remediation for sustainable management of arsenic in soil and water. *Environment International* 75, 180–198. DOI: 10.1016/j.envint.2014.11.010.

Saier Jr, M.H. (2016) Transport protein evolution deduced from analysis of sequence, topology and structure. *Current Opinion in Structural Biology* 38, 9–17. DOI: 10.1016/j.sbi.2016.05.001.

Salam, M., Varma, A., Chaudhary, D. and Aggarwal, H. (2020) Novel arsenic resistant bacterium *Sporosarcina luteola* M10 having potential bioremediation properties. *Journal of Microbiology & Experimentation* 8(6), 213–218. DOI: 10.15406/jmen.2020.08.00311.

Seiler, C. and Berendonk, T.U. (2012) Heavy metal driven co-selection of antibiotic resistance in soil and water bodies impacted by agriculture and aquaculture. *Frontiers in Microbiology* 3, 399. DOI: 10.3389/fmicb.2012.00399.

Sheng, G.-P., Xu, J., Luo, H.-W., Li, W.-W., Li, W.-H. *et al.* (2013) Thermodynamic analysis on the binding of heavy metals onto extracellular polymeric substances (EPS) of activated sludge. *Water Research* 47(2), 607–614. DOI: 10.1016/j.watres.2012.10.037.

Sheng, X.F., Xia, J.J., Jiang, C.Y., He, L.Y. and Qian, M. (2008) Characterization of heavy metal-resistant endophytic bacteria from rape (*Brassica napus*) roots and their potential in promoting the growth and lead accumulation of rape. *Environmental Pollution* 156(3), 1164–1170. DOI: 10.1016/j.envpol.2008.04.007.

Shofia, S.I., Vickram, A.S., Saravanan, A., Deivayanai, V.C. and Yaashikaa, P.R. (2025) Sustainable separation technologies for heavy metal removal from wastewater: An upgraded review of physicochemical methods and its advancements. *Sustainable Chemistry for the Environment* 10, 100264. DOI: 10.1016/j.scenv.2025.100264.

Silver, S. and Phung, L.T. (2005) Genes and enzymes involved in bacterial oxidation and reduction of inorganic arsenic. *Applied and Environmental Microbiology* 71(2), 599–608. DOI: 10.1128/AEM.71.2.599-608.2005.

Sujiritha, P.B., Vikash, V.L., Ponesakki, G., Ayyadurai, N. and Kamini, N.R. (2024) Microbially induced carbonate precipitation with arthrobacter creatinolyticus: An eco-friendly strategy for mitigation of chromium contamination. *Journal of Environmental Management* 365, 121300. DOI: 10.1016/j.jenvman.2024.121300.

Tarfeen, N., Nisa, K.U., Hamid, B., Bashir, Z., Yatoo, A.M. *et al.* (2022) Microbial remediation: A promising tool for reclamation of contaminated sites with special emphasis on heavy metal and pesticide pollution: A review. *Processes* 10(7), 1358. DOI: 10.3390/pr10071358.

Torres, E. (2020) Biosorption: A review of the latest advances. *Processes* 8(12), 1584. DOI: 10.3390/pr8121584.

Umrania, V.V. (2006) Bioremediation of toxic heavy metals using acidothermophilic autotrophes. *Bioresource Technology* 97(10), 1237–1242. DOI: 10.1016/j.biortech.2005.04.048.

Veglio, F. and Beolchini, F. (1997) Removal of metals by biosorption: A review. *Hydrometallurgy* 44(3), 301–316. DOI: 10.1016/S0304-386X(96)00059-X.

Vélez, J.M.B., Martínez, J.G., Ospina, J.T. and Agudelo, S.O. (2021) Bioremediation potential of pseudomonas genus isolates from residual water, capable of tolerating lead through mechanisms of exopolysaccharide production and biosorption. *Biotechnology Reports (Amsterdam, Netherlands)* 32, e00685. DOI: 10.1016/j.btre.2021.e00685.

Vidali, M. (2001) Bioremediation. An overview. *Pure and Applied Chemistry* 73(7), 1163–1172. DOI: 10.1351/pac200173071163.

Wróbel, M., Śliwakowski, W., Kowalczyk, P., Kramkowski, K. and Dobrzyński, J. (2023) Bioremediation of heavy metals by the genus *Bacillus*. *International*

Journal of Environmental Research and Public Health 20(6), 4964. DOI: 10.3390/ijerph20064964.

Xue, Y., Li, Y., Li, X., Zheng, J., Hua, D. *et al.* (2024) Arsenic bioremediation in mining wastewater by controllable genetically modified bacteria with biochar. *Environmental Technology & Innovation* 33, 103514. DOI: 10.1016/j.eti.2023.103514.

Yang, H.C., Fu, H.L., Lin, Y.F. and Rosen, B.P. (2012) Pathways of arsenic uptake and efflux. In: *Current Topics in Membranes*, Vol. 69. Academic Press. pp. 325–358.

Yang, K., Bu, H., Zhang, Y., Yu, H., Huang, S., Ke, L. and Hong, P. (2022) Efficacy of simultaneous hexavalent chromium biosorption and nitrogen removal by the aerobic denitrifying bacterium *Pseudomonas stutzeri* YC-34 from chromium-rich wastewater. *Frontiers in Microbiology* 13, 961815.

Yin, K., Lv, M., Wang, Q., Wu, Y., Liao, C. *et al.* (2016) Simultaneous bioremediation and biodetection of mercury ion through surface display of carboxylesterase E2 from *Pseudomonas aeruginosa* PA1. *Water Research* 103, 383–390. DOI: 10.1016/j.watres.2016.07.053.

Yue, Z.-B., Li, Q., Li, C. -c, Chen, T. -h and Wang, J. (2015) Component analysis and heavy metal adsorption ability of extracellular polymeric substances (EPS) from sulfate reducing bacteria. *Bioresource Technology* 194, 399–402. DOI: 10.1016/j.biortech.2015.07.042.

Zahran, E., Mamdouh, A.-Z., Elbahnaswy, S., El-Son, M.M.A., Risha, E. *et al.* (2025) The impact of heavy metal pollution: Bioaccumulation, oxidative stress, and histopathological alterations in fish across diverse habitats. *Aquaculture International* 33(5), 371. DOI: 10.1007/s10499-025-02045-1.

Zhang, C., Singh, R.P., Yadav, P., Kumar, I., Kaushik, A. *et al.* (2025) Recent advances in biotechnology and bioengineering for efficient microalgal biofuel production. *Fuel Processing Technology* 270, 108199. DOI: 10.1016/j.fuproc.2025.108199.

Zhang, W., Chen, L. and Liu, D. (2012) Characterization of a marine-isolated mercury-resistant *Pseudomonas putida* strain SP1 and its potential application in marine mercury reduction. *Applied Microbiology and Biotechnology* 93(3), 1305–1314. DOI: 10.1007/s00253-011-3454-5.

Zhou, Q., Pu, Y., Deng, H., Gong, J., Guo, L. *et al.* (2025) *Rhodobacter sphaeroides* reduces Pb accumulation by reshaping the intestinal microenvironment and improving liver oxidant resistance in common carp (*Cyprinus carpio* L.). *Journal of Hazardous Materials* 492, 138152. DOI: 10.1016/j.jhazmat.2025.138152.

Fungi-based Methods for Remediation of Heavy Metals from Polluted Sites: Recent Updates

3

3.1 Introduction

Industrialization and urbanization have directly affected the environment. Effluents and discharges from industry are the biggest source of pollution. The persistence of heavy metals (HMs) in these effluents imposes a great threat to aquatic life and human health through soil and aqueous streams. Such HMs are bio-persistent in nature and their toxicity is well cited in literature, especially in human beings, causing lethal health problems such as reproductive failure and possibly even infertility, kidney damage, shortness of breath, lung oedema, cancer development as well as psychological disorders. Expeditious urbanization and industrialization has undoubtedly led to exceptional growth but also has drastically affected the environment. As a result, the contamination and degradation of the whole ecosystem has become a major problem and a threat to all life forms, especially to human beings (Singh *et al.*, 2010). Increased availability of HMs such as cadmium (Cd), lead (Pb), mercury (Hg), zinc (Zn), chromium (Cr), arsenic (As) and copper (Cu) in natural resources is hazardous for life on earth. These HMs are extremely toxic and mainly contaminate the soil and water (Laoye *et al.*, 2025). Plants that grow in polluted soil can easily take up HMs that may then enter the food chain. Aquatic biota are badly affected by the presence of HMs. These HMs are released into the environment by natural and anthropogenic activities (El-Sharkawy *et al.*, 2025), for instance through forest fires, weathering of rocks, etc. Rocks are the natural sources of HMs, and magma is the molten rock that consists of chemical elements due to plate tectonics and volcanism. Physical damage to rocks due to bad weather conditions leads to the incorporation of HMs into them. The various HMs present in rocks include Cd, nickel (Ni), molybdenum (Mo), selenium (Se), vanadium (V), rubidium (Rb), cobalt (Co), manganese (Mn), lithium (Li), Zn, gallium (Ga), strontium (Sr), fluorine (F), Cu, barium (Ba),

Corresponding author: sukhminderjit.uibt@cumail.in

Pb, etc. (Alengebawy *et al.*, 2021). They are found in the form of sulfides and oxides of metals. In soils, they are found as sulfides of metals in combination with sulfides of copper and iron. HMs are released as by-products of various hydro-metallurgical processes after mining (El-Sharkawy *et al.*, 2025).

Many industries release HMs in their effluents which enter the soil and thus into food chains by absorption through roots. Paints, preservatives, insecticides and phosphate fertilizers, etc., also emit HMs and these to accumulate in ecosystems causing serious threat to the environment (Hu *et al.*, 2016). Soil can adsorb, exchange, oxidize, catalyse and reduce metal ions. From the surface of soil, HMs ultimately find their way into sewage and reservoirs. Metal wastes, wastewater irrigation, animal manure and sludge can contaminate the soil with metals and metalloids. Industrial wastes and landfill leachates can contaminate groundwater. Volcanic eruptions, mineral dust and seasalt particles can act as sources of HMs in the atmosphere. There are many human-induced sources of HMs which mainly include sewage sludge, paper industries, pesticides, batteries, tanneries, fertilizer industries and wastewater irrigation (Briffa *et al.*, 2020). HMs can be released in compound as well as in elemental forms. Pesticides and fertilizers that are used for crop production also introduce HMs to the environment.

Researchers have explored various strategies to remove HM contaminants and to restore environmental quality. Among these, various physical remediation methods and chemical remediation treatments have been widely used but these approaches present several drawbacks such as they are highly expensive and not very effective (Kumar *et al.*, 2021). The most economical and eco-friendly way to treat this pollution is through bioremediation where microbes are employed to transform or convert HM toxins into less harmful states using various mechanisms (Maglione *et al.*, 2024). Bioremediation can also produce toxic microbial metabolites that can interfere in the whole process of remediation, which leads to non-degradation of HMs. However, in comparison with other physical and chemical methods, this technique is a better choice in terms of its eco-friendly nature as well as cost-effectiveness (Kondakindi *et al.*, 2024).

According to the US Environmental Protection Agency (EPA), bioremediation is a technique that uses naturally occurring microorganisms to break down hazardous substances into less toxic or non-toxic substances. Bioremediation offers an economical approach to remove these contaminants from the environment and uses both live microorganisms as well as plants (phytoremediation) for the removal of HMs. Many microorganisms have developed different resistant mechanisms for HMs (Kuppan *et al.*, 2024). This is the reason that microbial technologies are gaining the interest of scientists for HM removal. Many of these microbes can be successfully utilized at contaminated sites (Pande *et al.*, 2022).

Among the wide range of microorganisms used for bioremediation, there are many fungal strains used for this purpose. Mycoremediation is a biological process that employs fungi, either living or dead, to remove toxic substances from the environment. Mycoremediation is an eco-friendly approach and holds potential in removing toxic HMs from the environment by various mechanisms such as transformation, degradation or immobilization without the generation of any hazardous by-products. Various types of fungi have been used in bioremediation of soil and aquatic ecosystems and its success largely depends on choosing the right fungal species for specific contaminants (Dinakarkumar *et al.*, 2024).

3.2 Fungi in Heavy Metal Remediation

Fungi represent a diverse genera among microorganisms composed of many identified species, with only a small fraction being explored for their mycoremediation potential. Fungi are versatile genera and have evolved to survive and function effectively under diverse environmental stresses and extreme conditions, making them promising agents for bioremediation. Fungi transform, degrade or immobilize pollutants like heavy metals, hydrocarbons and pesticides efficiently by producing various secondary metabolites such as polyketides, terpenoids, alkaloids and phenolics, which play an important role in detoxification mechanisms (Roy *et al.*, 2025). Different fungi such as white rot and filamentous fungi have been explored in different studies for HM removal in both soil and aquatic systems. Different genera belonging to basidiomycetes/white rot fungi such as *Pleurotus ostreatus*, *Phanerochaete chrysosporium* and *Trametes versicolor* have been explored for HM detoxification. The fungi have shown potential to remediate HM-contaminated environments through different mechanisms like biosorption, extracellular precipitation, bioaccumulation, enzymatic transformation and production of biosurfactants (Singh and Singh, 2024). Ascomycetes, also known as filamentous fungi, have shown tolerance to high concentrations of the HMs Pb, Cd, Cr and Ni, and promise in their removal of metals from wastewater. Various filamentous fungal species of *Aspergillus*, *Penicillium* and *Trichoderma*, and black yeasts like *Exophiala*, produce fast-growing filamentous forms and bind and immobilize metals efficiently onto their cell walls (Iram *et al.*, 2012; Zareh *et al.*, 2022; Dagdag *et al.*, 2023).

Different ecological groups of fungi, including saprophytic, mycorrhizal, yeast/dimorphic and endophytic or extremophilic forms, contribute to metal detoxification through mechanisms such as biosorption, bioaccumulation, enzymatic transformation and biomineralization (Table 3.1). Thus, fungi offer a biologically robust and scalable solution for HM remediation, with current research focusing on optimizing conditions, exploring genetic improvements and field application for complex industrial and environmental contamination (Paria *et al.*, 2022). Saprophytic fungi are well studied for their extracellular enzyme systems and robust cell wall interactions, whereas mycorrhizal fungi enhance phytoremediation by facilitating metal uptake and immobilization

Table 3.1. Different fungal species showing heavy metal (HM) detoxification.

Fungal type	Fungal species	Heavy metals	Outcome	Reference
Saprophytic fungi	White rot fungi (*Phlebia brevispora* and *P. floridensis*)	Ni, Cd, Pb	Maximum removal Ni (99–98%), Cd (98–97%) and Pb (12–98%) from industrial wastewater depending on fungal species used	Sharma *et al.*, 2023
Saprophytic fungi	*Pleurotus* sp.	Co, Cu	Increased activity of superoxide dismutases and catalase observed along with increased accumulation of HM in mycelium	Mohamadhasani and Rahimi, 2022
Saprophytic fungi	*Cladosporium* sp. (multi-metal hyper-tolerant fungus)	Multi-metals	Dead biomass of fungus showed high adsorption of HMs such as Cr, Cd, Pb, Ag, Cu, Ni, Mn, Zn, Fe, Al and Co in industrial effluents under optimized operating conditions	El-Gendy *et al.*, 2023
Mixed fungi	Dead mixed fungal mycelium	Pb	Dried mycelium membrane demonstrated high Pb(II) removal efficiency up to 85–90% at the highest Pb(II) concentrations	Parasnis *et al.*, 2024
Saprophytic fungi	*Aspergillus flavus* biomass	Cs^+ and Sr^{2+} ions	Demonstrated significant uptake of HMs and holds potential as promising, low-cost biosorbent for nuclear wastewater treatment	Mousa *et al.*, 2025

Continued

Table 3.1. Continued

Fungal type	Fungal species	Heavy metals	Outcome	Reference
Arbuscular mycorrhizal fungi (AMF)	*Glomus monosporum, G clarum, Gigaspora nigra* and *Acaulospora laevis*	Cd	Higher tolerance against Cd stress in AMF-inoculated plants with high production of antioxidant enzymes and malondialdehyde content	Abdelhameed and Metwally, 2019
Yeast	*Geotrichum* sp.	Multi-metals	Accumulated high levels of HMs especially Cu^{2+}, Zn^{2+} and Ni^{2+} and removed efficiently from polluted water	He *et al.*, 2022

in plants. Yeasts and dimorphic fungi provide resilience under fluctuating environmental conditions, while endophytic and extremophilic fungi exhibit remarkable tolerance to toxic environments, allowing remediation in extreme habitats. This functional and ecological diversity underscores the potential of fungi as versatile and sustainable tools for bioremediation of metal-polluted ecosystems (Dinakarkumar *et al.*, 2024).

3.2.1 Saprophytic fungi

Saprophytic fungi are the most extensively explored fungi as workhorses for mycoremediation. They include filamentous ascomycetes (e.g. *Aspergillus, Penicillium*), mucoromycetes (e.g. *Rhizopus*) and wood-decaying basidiomycetes (white, brown and soft rot; e.g. *Pleurotus, Trametes, Phanerochaete*). These fungi can decrease the bioavailability of heavy metals, as reported in the literature, owing to the presence of chemically versatile cell walls (chitin–β-glucan with carboxyl, phosphate, amine and thiol groups), abundant extracellular polymeric substances (EPS), pigments (notably melanin) and low-molecular-weight organic acids. These fungi demonstrate biosorption, bioaccumulation, redox transformation and biomineralization, enabling removal of Pb, Cd, Cu, Zn and iron (Fe) from contaminated matrices. White rot fungi demonstrate considerable potential for sustainable application in remediating HM-contaminated environments. *Phanerochaete chrysosporium, Pleurotus ostreatus* and *Trametes versicolor* have been extensively studied for HM remediation (Chen *et al.*, 2022). With increasing concentrations of substrate, *P. ostreatus* showed higher accumulations of HMs such as Cu, As, Cd and Pb (Atila and Kazankaya, 2023).

3.2.2   Mycorrhizal fungi

Mycorrhizal fungi form symbiotic relationships with plant roots, mutually ben-efiting each other, where the plant supplies carbon (C) while the fungi enhance water and nutrient uptake through the roots. Mycorrhiza fungi are further classified on the basis of how they colonize the plant root as either arbuscular mycorrhizal fungi (AMF), where the fungi penetrate the plant root cells, and ectomycorrhizal (ECM) fungi, where the fungi inhabit the surface of the root. Mycorrhizal fungi play a pivotal role in remediating HM-contaminated soils through symbiotic interactions with their host plants. These fungi enhance plant growth, reduce metal toxicity and stabilize metals in the rhizosphere via sequestration, chelation and mineral precipitation. With hundreds of identified species, they are present in nearly all natural and agricultural ecosystems, forming associations with the vast majority of plant species. In fact, it is estimated that around 90% of terrestrial plants coexist with AMF. About 240 different species of AMF belong to the Glomeromycota subfamily, which are found in almost all natural and agricultural ecosystems and coexist with most plants (Parvin *et al.*, 2019). AMF form intracellular arbuscules and vesicles within root cortical cells, expanding the absorptive surface area. Their hyphae accumulate HMs in cell walls and vacuoles, while glomalin and other extracellular glycoproteins bind and immobilize metals in soils. ECM fungi, with extensive extraradical mycelia, enhance nutrient exchange while secret-ing organic acids and siderophores that alter metal solubility. Both groups of mycorrhiza reduce the translocation of toxic metals to plants and hence reduce their availability in food chains (Lanfranco and Young, 2012).

Various studies have explored the role of ectomycorrhiza in mycoreme-diation. In one study, Zhang *et al.* (2025) studied the potential of the ecto-mycorrhizal fungus *Suillus luteus* along with biochar in *Pinus massoniana*. This study highlighted the synergistic roles of fungi, plants and microbes in the detoxification of As by immobilization, aggregate formation and soil carbon stabilization. Another study reported improved Pb tolerance with AMF inoculation in *Medicago truncatula* plants by strengthening cell wall structures and enhancing Pb immobilization. Inoculation with AMF showed increased Pb accumulation in cell walls along with higher biomass, elevated polysaccharide levels in pectin and hemicellulose fractions and higher cell wall peroxidase activity in AMF-associated roots under Pb stress (Zhang *et al.*, 2021).

Arbuscular mycorrhizal fungi play a crucial role in reducing Cd toxicity in plants. AMF mitigate Cd stress by limiting Cd uptake and transport, enhancing nutrient status, boosting antioxidant defense and maintaining ion balance. They also influence Cd speciation and compartmentalization, and regulate metal transporter genes along with phytochelatin and metallothionein syn-thesis. The effectiveness of AMF, however, varies with soil conditions, plant species and fungal taxa, highlighting their potential in specific phytoremedia-tion and Cd-tolerant crop development (Zhao *et al.*, 2024).

3.2.3 Yeasts and dimorphic fungi

Yeasts have emerged as promising candidates for the bioremediation of HM-contaminated environments. Their unique metabolic pathways, adaptability to diverse conditions and ability to transform or sequester toxic metals make them valuable tools in environmental clean-up efforts. Yeasts are widely used as model microorganisms due to their easy cultivation, rapid growth and genetic stability. Strains like *Saccharomyces cerevisiae* and *Rhodotorula mucilaginosa* show strong capacities for HM and metalloid removal through mechanisms involving their cell walls, intracellular proteins and EPS. Functional groups in the cell wall adsorb metals, intracellular proteins aid in detoxification and EPSs facilitate metal complexation. Due to these abilities, yeasts serve as effective bioremediation agents and can be combined with other materials to enhance metal removal (Shao *et al.*, 2025).

A novel yeast, *Geotrichum* sp. CS-67, tolerant to heavy metals and screened from polluted mangrove soils, showed strong tolerance to Cu^{2+}, Zn^{2+} and Ni^{2+} suggesting a potent candidate for microbial remediation of HM contamination. The yeast sequestered Zn^{2+} most efficiently, followed by Ni^{2+} and Cu^{2+}, with Zn^{2+} and Ni^{2+} actively taken up by cells and Cu^{2+} primarily adsorbed to the cell wall. RNA-Seq analysis revealed differential expression of numerous genes involved in HM processing, and quantitative reverse transcription polymerase chain reaction (qRT-PCR) validation showed upregulation of *SED1* and *GDI1* and downregulation of *ZRT1* under Zn^{2+} and Ni^{2+} stress (He *et al.*, 2022).

Endophytic and extremophilic fungi have garnered significant attention in recent years for their potential in bioremediation of HMs. Endophytes, residing within plant tissues without causing harm, can sequester, transform and detoxify heavy metals (Bhardwaj, 2025).

3.3 Mechanisms of Metal Remediation by Fungal Species

Remediation of HMs in the environment by fungi is carried out using a wide range of mechanisms whereby physicochemical and biological processes are used collectively to effectively detoxify the environment. Such strategies involve: (i) biosorption, in which the metal ions are passively adsorbed by the fungi cell walls; (ii) bioaccumulation, in which the metal is actively sequestered and complexed within the fungi cell walls using a complex of biomolecules; (iii) enzymatic transformation, in which fungal enzymes oxidize, reduce and methylate metal toxicity; and (iv) efflux systems to lose surplus metal ions, thus preserving intracellular homeostasis and promoting biomineralization and precipitation, reducing soluble metals to insoluble forms, including oxalates and carbonates (Fig. 3.1). The combination of these synergistic processes allows fungi to survive in metal-rich environments, and is critical in bioremediation.

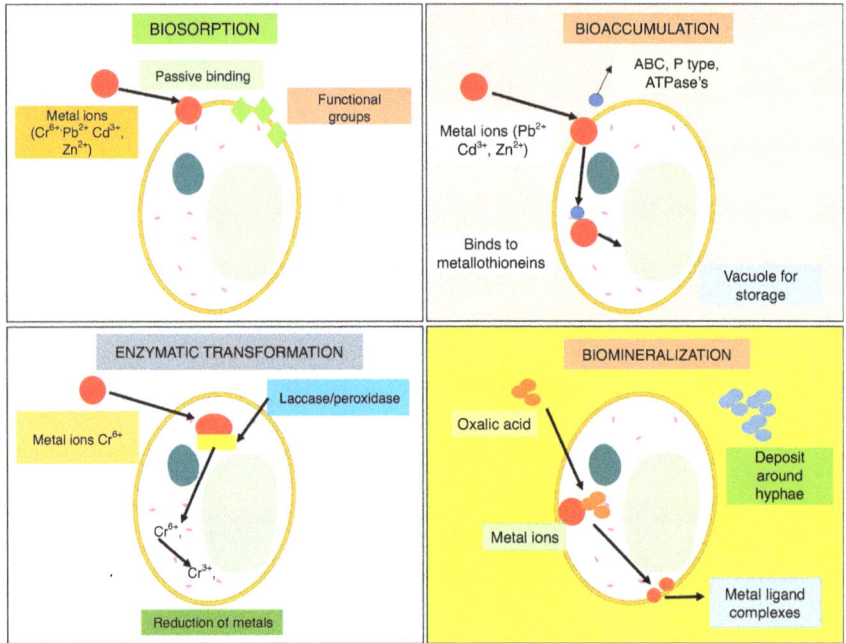

Fig. 3.1. Mechanisms of metal remediation by fungal species.

3.3.1   Biosorption and cell wall interactions

Biosorption is the dominant first-contact mechanism, involving passive binding of metal ions to cell wall/EPS ligands. Dried biomass often matches or surpasses living biomass in capacity, while avoiding nutrient demands. Fungal cell walls, primarily composed of chitin, glucans and proteins, play a pivotal role in the biosorption of HMs. These structures possess functional groups such as hydroxyl, carboxyl, amino and phosphate groups, which facilitate the binding of metal ions through various mechanisms. In adsorption, the metal ions present in the environment will bind to the fungal cell wall through non-specific interactions such as van der Waals forces and electrostatic attractions. This process is often pH-dependent and can be influenced by the surface charge of the fungal cells. The fungal cell walls contain functional groups that can exchange metal ions with protons or other cations present in the environment. This ion-exchange mechanism contributes to the removal of HMs from contaminated sites. Additionally, fungi can produce metabolites such as organic acids and siderophores that bind to metal ions, forming stable complexes. These complexes reduce the toxicity and mobility of the metals, facilitating their sequestration. Fungal cells can induce the precipitation of metal ions by altering the local pH or through the production of metabolites that promote the formation of insoluble metal compounds. This process effectively immobilizes the metals within the fungal biomass.

Aspergillus niger combined with fluorapatite effectively immobilizes Pb, with stability strongly influenced by pH. *A. niger* grows well between pH 2.5 and 6.5, and over 99% of Pb is removed at pH 3.5–6.5, with minimal re-release at pH 3.5–5.5. Extremely acidic conditions (pH 1.5) hinder fungal growth and reduce Pb removal. Pb remediation is mainly driven by lead oxalate and lead/calcium oxalate formation, while higher pH (>5.5) lowers mineral stability. This combination shows great potential for Pb remediation in mildly acidic environments (Feng *et al.*, 2025).

Resource recovery from precious metal-laden wastewater is vital for sustainable development, with superalloy electrolytes offering significant recovery potential. In this study, five metal-tolerant fungi were screened, and *Paecilomyces lilacinus* exhibited the highest adsorption capacity for Co, Cr, Mo, rhenium (Re) and Ni under optimized conditions. Adsorption modelling indicated a multi-step process dominated by chemical adsorption, with active groups such as hydroxyl, amino, carbonyl, carboxyl and phosphate involved. These findings highlight the promise of *P. lilacinus* for treating complex industrial wastewaters containing multiple HMs (Yang *et al.*, 2024).

3.3.2 Bioaccumulation and intracellular sequestration

Fungi possess remarkable abilities to tolerate and remediate HM contaminants through bioaccumulation and intracellular sequestration mechanisms. Bioaccumulation involves intracellular detoxification that involves the uptake of metal ions into fungal cells, followed by their sequestration in vacuoles or binding to intracellular biomolecules, thereby reducing metal toxicity and facilitating detoxification. Fungi utilize various transporters, including ATP-binding cassette (ABC) transporters and P-type ATPases, to uptake metal ions from the environment. Fungi generally withstand elevated intracellular metal concentrations through two main strategies: chelation and compartmentalization. In the chelation process, metal-binding ligands such as metallothioneins, phytochelatins and glutathione are synthesized and bind HMs within the cytosol, primarily via their reactive thiol (-SH) groups, and reduce metal toxicity. In compartmentalization, surplus metal ions are sequestered into intracellular organelles such as vacuoles through polyphosphate-mediated transport, effectively reducing their cytotoxic impact (Shao *et al.*, 2025). The vacuole serves as a primary compartment for storing excess metals. Fungal cells may sequester metals in vacuoles through processes like endocytosis and vesicular trafficking, effectively isolating them from cytoplasmic components and reducing their bio-availability. These complexes facilitate the detoxification of metals by rendering them less reactive and less likely to interfere with cellular processes. Energy-dispersive X-ray spectroscopy confirmed that Pb bioaccumulation occurred predominantly within the vacuoles in *Rhodotorula mucilaginosa* YR29 (Angeles de Paz *et al.*, 2023). Khandelwal *et al.* (2022) reported that yeast cadmium factor 1 (Ycf1), an ATP-binding cassette C-subfamily type transporter, sequesters HMs and glutathione into vacuoles to mitigate cellular stress in yeast.

3.3.3   Enzymatic transformation and detoxification

Fungi produce many enzymes to detoxify HMs in contaminated environments depending on the type of HM, environmental conditions and type of fungal strain. Exposure to HMs triggers the activation of various antioxidant defense enzymes, including catalase, superoxide dismutase and glutathione transferase. Catalase plays a vital role by converting hydrogen peroxide into water and oxygen, while superoxide dismutase protects the cell from reactive oxygen species (ROS). Additionally, several enzymes facilitate oxidation–reduction reactions of HMs, helping minimize cellular damage by reducing bioavailability. Fungal enzymes, particularly laccases and peroxidases, catalyse the oxidation of HMs, leading to the formation of less toxic or insoluble forms. For instance, laccases can oxidize metal ions, facilitating their precipitation and removal from the environment. The purified laccase from *Ganoderma multipileum* has shown over 94% reduction of hexavalent Cr(VI) into the less toxic trivalent Cr(III) form (Alshiekheid *et al.*, 2023). Other white rot fungi such as *Pleurotus florida* have been reported to biosorb and leach Cu and Fe from e-waste using laccase enzyme activity (Kaur *et al.*, 2022). The expression of various enzymes to laccase, lignin peroxidase and Mn peroxidase and the production of organic acids by *Phanerochaete chrysosporium* facilitated 54% copper leaching (Liu *et al.*, 2022). Furthermore, its role in iron leaching (18% from pyrite) was linked to enzymes, hydrogen peroxide and organic acids (Yang *et al.*, 2018), and the involvement of cytochrome P450 (CYP450) in Cd sequestration was confirmed through inhibitor studies (Zhang *et al.*, 2025). Beyond individual metals, *P. chrysosp*orium has also been implicated in the removal of multiple toxic HMs including As, Zn, Pb, Cr, Co and Ni (Rudakiya *et al.*, 2018; Akhtar and Mannan, 2020). Laccases and peroxidases are the main enzymes produced by fungal species in HM-contaminated sites (Singh and Singh, 2024).

3.3.4   Efflux systems

Microorganisms can remove pollutants from their cells by the efflux system – the process of expelling pollutants out of their cells. Fungi also use different efflux systems to remove HMs out of their cells to reduce the intracellular concentration of HMs. Different types of efflux systems have been characterized in fungi and some of them are discussed here.

The P-type ATPases are membrane-bound ATP-dependent transporters that are involved in actively expelling the toxic HM ions from the cytoplasm or channelizing them to intracellular organelles such as vacuoles and hence lowering their concentration in the cytoplasm. This efflux system in fungi uses ATP hydrolysis to export HMs across membranes. For example, CrpA is a P1B-type ATPase exporter, which is associated with transport of Cu and Cd ions out of the cytoplasm, thereby protecting fungi such as *Aspergillus fumigatus* and *A. nidulans* from HM toxicity (Vig *et al.*, 2023).

The efflux system using cation diffusion facilitators (CDFs) to maintain HM haemostasis in the fungal cytoplasm does this by transporting HMs as divalent cations. HMs such as Ca^{2+}, Co^{2+}, Fe^{2+}, Mn^{2+} and Ni^{2+} are either pumped out of the cytoplasm by proton/cation antiporters or sequestered into the vacuoles to prevent cytoplasmic toxicity and ensure metal homeostasis. For example, the CDF family of zinc transporters plays a crucial role in Zn homeostasis in fungi. Two key vacuolar membrane transporters, Zrc1p and Cot1p, mediate Zn sequestration in the vacuole and play an important role in Zn haemostasis (MacDiarmid *et al.*, 2000).

Fungi also remove HMs from the cytoplasm by ABC transporters, which during ATP hydrolysis export HMs into the vacuoles for sequestration. It has been reported that in *Saccharomyces cerevisiae*, the Ycf1 efflux pump transports Cd glutathione–cadmium conjugates into the vacuole, thereby reducing free cytoplasmic Cd concentration (Li *et al.*, 1997). This mechanism highlights the integration of metal chelation and efflux in fungal detoxification.

The major facilitator superfamily (MFS) is another efflux system found in fungi in which the proton motive force is utilized for removing many toxic compounds from their cells, which can contribute indirectly to the export of HMs. For example, *Candida* sp. Can be employed for resistance against Cd and other toxic ions. Efflux systems for metalloid oxyanions, such as chromate (CrO_4^{2-}) and arsenite (AsO_2^-), have been reported in *Aspergillus* sp. and *Candida* sp. and are often mediated by aquaglyceroporins or transporters homologous to bacterial Ars systems (Cavalheiro *et al.*, 2018). Compared with bacteria, fungal efflux systems show some distinctive features. While both groups rely on P-type ATPases and CDFs, fungi make greater use of ABC transporters and also combine efflux with vacuolar sequestration, which is less common in bacteria. In contrast, bacteria often employ large tripartite resistance–nodulation–division (RND) type efflux complexes, which are absent in fungi. Thus, fungal efflux mechanisms represent an evolutionarily adapted strategy that balances metal efflux across the plasma membrane with intracellular compartmentalization, ensuring survival in metal-rich environments and contributing to their bioremediation potential (Buechel and Pinkett, 2020).

3.3.5   Precipitation and biomineralization

Fungi can transform soluble metal ions present in the soil or aquatic ecosystems through mineralization into insoluble mineral forms. Through biomineralization, the soluble metals are transformed into minerals and the environment is decontaminated by immobilizing these toxic metal. Various metabolic activities of fungi, such as organic acid production, act as chelating agents helping in the mineralization and precipitation of HMs. *Aspergillus niger* has been shown to produce oxalate, leading to the precipitation of metal ions

pH	Low pH → H⁺ competition → ↓ Metal binding Neutral–Alkaline pH → Deprotonation of sites → ↑ Metal uptake
Temperature	Optimum (25–35 °C) → ↑ Metabolism & enzyme activity → ↑ Uptake High/Low extremes → Protein denaturation & stress → ↓ Uptake
Initial Metal Concentration	Low conc. → More free binding sites → ↑ Uptake High conc. → Binding site saturation & toxicity → ↓ Uptake
Nutrient Availability	Balanced C/N → ↑ Enzyme secretion (oxidoreductases, EPS) → ↑ Detoxification Nutrient limitation → ↓ Enzyme activity → ↓ Uptake
Co-pollutants/ Mixed Systems	Competition → ↓ Uptake of individual metals Additive/Synergistic → ↑ Total metal removal

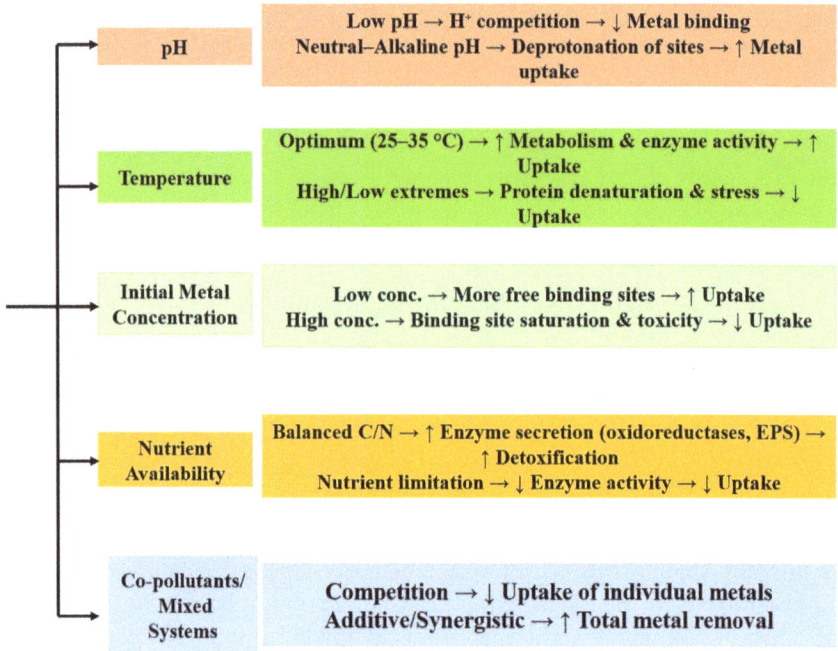

Fig. 3.2. Factors influencing fungal bioremediation efficiency.

like calcium and iron (Li *et al.*, 2024). Fungi can convert dissolved metal ions through various metabolic and enzymatic processes into mineral phases. This phenomenon of conversion of HMs into insoluble metal compounds by producing various extracellular enzymes is called precipitation. For instance, the white rot fungus *Phanerochaete chrysosporium* has been reported to precipitate metals such as Cu and Zn through the activity of laccase enzymes (Singh and Singh, 2024).

3.4   Factors Influencing Fungal Bioremediation Efficiency

The efficiency of fungal bioremediation depends on several physicochemical and biological factors (Fig. 3.2). For example pH, which is one of the most crucial factors for the biosorption process and relies on the ionization of functional groups such as the carboxyl, amino and phosphate groups present in the fungal cell wall.

For low pH, the carboxyl, amino and phosphate groups are mainly protonated and a strong competition between proton H⁺ and metal cations can be observed for binding sites, which negatively affects metal uptake efficiency. On the other hand, in the near-neutral to slightly alkaline pH range, the deprotonation of these side groups increases the negative charge on the fungi surface, which raises their metal-binding ability. For example, studies on *A. japonicus* showed the complete elimination of Ni(II) at pH 9 while that for Fe(II) had the highest removal at pH 6 (Rostami and Joodaki, 2002). Similarly, sorption of Pb and Cd by *Penicillium* sp. was found to be maximal at pH values of 6–7; above this range higher pHs favoured precipitation of the metal, limiting biosorption potential (Sánchez-Castellón *et al.*, 2022).

Temperature is another determining factor since it impacts fungal metabolism and enzyme kinetics, and the diffusion kinetics of the metal ions. At temperatures within an optimal range (25–35°C), increased temperature can improve sorption through the mobility of metal ions and biomass activity of the fungal biomass (Ramya *et al.*, 2021). In excess of this range, however, protein denaturation and metabolic stress depresses the efficiency of biosorption. *Trichoderma viride* and *Saprolegnia delica* demonstrated the highest ability to absorb metals at about 25°C, indicating that they slow down at temperatures above and below this value (Ali and Hashem, 2007). In the same manner, *A. niger* showed the most uptake of Cu(II) at 31°C and the performance diminished rapidly above 40°C (Dusengemungu *et al.*, 2020).

The availability of nutrients (mainly carbon and nitrogen) along with enzyme production are also contributing factors for bioremediation potential. The production of enzymes, organic acids, siderophores and polymers like EPS are dependent on the nutritional composition of the environment. Optimum availability of carbon and nitrogen sources enables the fungus to withstand HM exposure and enhance cellular detoxification. On the other hand, nutrient limitation may trigger enzyme production and excess glucose may result in catabolized repression by limiting the activity of ligninolytic enzymes and oxidoreductases, the enzymes mainly required in metal complexation and precipitation (Thepbandit and Athinuwat, 2024).

The initial concentration of HMs in the polluted medium also determines the productivity of fungal bioremediation. A high percentage removal is possible at low HM concentrations due to the high number of available active sites on the fungal cell walls. As the level of metals rises, however, binding sites are saturated and instead of being removed, the metal starts to exert toxic effects and fungal growth is inhibited. For example, in *Penicillium* sp. a removal of 59 mg/g was achieved up to an initial Pb concentration of 100 mg/l, but a higher concentration decreased removal efficiency (Ayele *et al.*, 2021). Similarly, *Pleurotus* sp. also grew exponentially in low concentration of Cu and Co but high concentrations resulted in severe reduction of their growth (Mohamadhasani and Rahimi, 2022).

The presence of other pollutants in the environment may synergistically or antagonistically affect fungal metal extraction. When more than one HM is present, the individual metals will compete for the binding sites which may decrease the uptake of one particular metal. An example is where *Mucor rouxii* cultivated in mixed-metal solutions had a lower rate of biosorption of individual metal ions but an increased total uptake capacity, so the effects were competitive but additive (Yan and Viraraghavan, 2003). Further, the presence of organic pollutants or other stress factors in industrial effluents may affect fungal cell surface or enzyme activity, and affect remediation favourably or unfavourably. These interactions highlight the intricacy of the *in situ* remediation conditions in the field – multi-pollutant setups necessitate mobile fungal responses to support successful bioremediation in the natural world (Bhattacharya *et al.*, 2020).

3.5   Recent Advances and Biotechnological Interventions

Recent studies are exploring the application of newer biotechnological approaches in mycoremediation with significant progress. Although still in infancy, genetically engineered fungi are being developed with higher efficiency in HM removal. Elahian *et al.* (2020) engineered a metal-resistant yeast, *Pichia pastoris*, strain for bioleaching and reported higher biosorption efficiency for gold and palladium. Applications of immobilization techniques are also being explored for combating HM pollution and to enhance the practicability of fungal biomass in wastewater treatment. Immobilized fungal strains like *Penicillium janthinillum* in polyvinyl alcohol-sodium alginate beads have shown huge promise. Immobilization strengthens the mechanical strength, stability and recycling of the adsorbent, and keeps the biosorption capacity of metals, such as Pb, Cd and Cu, at an appreciable level (Cai *et al.*, 2016). Moreover, natural support materials such as loofa sponge are considered to be an economical and non-toxic platform of immobilization that enhances high surface area and support. The use of an alginate-based bead is broad because it has a carboxyl-rich composition, which facilitates ion exchange capability and the binding of HMs, thereby making the bead suitable for use in repeated cycles (Ali *et al.*, 2021).

Nanotechnology is another promising fungal bioremediation field. Production of nanoparticles including silver, gold and zinc oxide as well as cadmium sulfide can be achieved using fungi as a biofactory source, providing an alternative and clean substitute to chemical production (Rezghi Rami *et al.*, 2024). Different nanoparticles synthesized using fungi have shown significant removal of HMs present in contaminated soil and water bodies (Sidhu *et al.*, 2025). Hybrid biosorbents prepared by incorporation of fungal biomass with nanoparticles increases biosorption capacities and catalytic activity which can be useful in removing complex pollutants, such as dyes and HM mixtures.

Integrating different omics approaches, such as genomics, transcriptomics, metabolomics and proteomics, is further enhancing the understanding of the mechanisms of fungal responses to HMs. An extensive shift in gene expression has been demonstrated by transcriptomes studies which have found increased expression of oxidative stress responsive genes and ABC transporters genes involved in metal sequestration (Tian *et al.*, 2024). Similarly, proteomic studies have reported the prevalence of various stress-response, detoxification and antioxidant proteins that are critical in keeping the cell in homeostasis during HM toxicity. These insights are crucial for guiding metabolic engineering approaches for strain development with enhanced bioremediation potential (Chiapello *et al.*, 2015).

3.6   Field Applications and Case Studies

Mycoremediation has been developed beyond the laboratory, and is now being applied on industrial scale for waste treatment and cleaning up of mining-affected sites. A significant level of HM removal has been reported using *Aspergillus fumigatus* for bioleaching mine tailings. In column experiments, *A. fumigatus* removed HMs such as As (53%), Fe (51%), Mn (81%) and Zn (62%) by producing different organic acids, especially oxalic acid, produced in the process of metabolism. This remediation capacity of *A. fumigatus* on such complex matrices explains its potential in large-scale remediation of mine-camped materials (Seh-Bardan *et al.*, 2012). *A. niger* is highly efficient at solubilizing Cu in mining tailings via organic acid-mediated leaching with removal efficiencies up to 63% using sucrose as a carbon source and 80% when ultrasound preconditioning was included in the procedure. These results show that it is possible to industrialize fungal bioleaching to clean up Cu-loaded environments on the industrial scale.

In HM-contaminated soil, fungi have been also used in remedying/treating the persistent organic pollutant (e.g. polychlorinated biphenyls (PCBs)) (Mulligan and Galvez-Cloutier, 2003). A smaller study was carried out in a pilot-scale reactor with spent mushroom substrate of *Pleurotus ostreatus* to decontaminate 10 tons of PCB-contaminated soil. A reduction of 94.1% in PCBs was recorded after 8 months of treatment when working with this system. The oxidative-promoting effect of fungal activities, along with bacteria metabolism and the action of enzymes, was associated with this degradation (Siracusa *et al.*, 2017). Related field experiments using *P. ostreatus* have indicated rates of PCB biodegradation of up to 50.5% in the rhizosphere region and around 18.5% within bulk soil in 12 weeks, as well as stimulated growth of the native microbial community in the host environment and increased degradation rates in general (Stella *et al.*, 2017).

Fungi have also been promising as industrial wastewater treatment systems. Air-lift and fluidized types of bioreactors have been used to treat hospital wastes containing pharmaceutical pollutants in the presence of white rot fungi like *Trametes versicolor*. Such systems have been shown to remove up to 80% of pharmaceutical active compounds over several days of running time. Also, rotating drum bioreactors with immobilized *T. versicolor* pellets have been examined with pesticide-contaminated agricultural wastewater with a removal efficiency of about 87% (Amobonye *et al.*, 2023). These bioreactor systems not only enhance pollutant degradation by the fungus, but also allow fungal growth over extended timeframes of operation which makes them a promising technology to apply at industrial scales.

These examples depict the reliability and malleability of fungi in the context of extensive remediation projects. Fungi have been applied in a wide range of environments including mining tailings, e-waste-contaminated soils and complex industrial effluents and can be scaled to offer environmentally friendly pollution control solutions (Beltrán-Flores *et al.*, 2023).

3.7   Challenges and Limitations

Despite studies supporting the utilization of fungal bioremediation, there are practical and biological factors inhibit its implementation in the real world. Among these, one drawback is the slow growth rate of fungi in comparison to that of bacteria. Although fungal metabolism is an important aspect of biosorption and detoxification, decreased rates of biomass production translate to the possible need to leave processes for longer and for potentially higher volumes of fungal inoculum in time-sensitive or high-throughput applications (El-Gendi *et al.*, 2021). Metal recovery after biosorption by fungal mycelium is another challenge. Fungi have the ability to readily collect HMs but it is not that easy to obtain the HMs in reusable or original forms. Biomineralization and intracellular entrapment add complexity to the downstream phase of recovery and fungal biomass may be damaged by desorption methods and the process may not be efficient, limiting the practicality of recycling (Ramya *et al.*, 2021). Variation in quality of performance in the case of varying environmental conditions is another impediment. Removal efficiencies can strongly depend on such factors as pH changes, excessively high or low temperatures, nutrient sources and inorganic ions in water or soil. The results obtained in laboratories under controlled conditions often have a higher success rate than can be realized in reality because environmental heterogeneity does not allow reproducibility and results in low reliability in various regions or over time (Fouda-Mbanga *et al.*, 2024).

The other issue that is usually not considered is the danger of pathogenicity which characterizes some fungal species. Although most of the studies on bioremediation deal with non-pathogenic strains, it is important to frame the concept of environmental safety in view of opportunistic pathogens such as *A. fumigatus* and *A. flavus*. Any massive emission or concentration of such species in polluted places presents safety and ecological concerns (Corbu *et al.*, 2023).

3.8   Future Prospects

Fungal bioremediation can also be improved by combining it with other bioremediation approaches such as synergy with plants and bacteria. As an example, mycorrhizal-assisted phytoremediation – a process of forming symbiosis between AMF and plants – has demonstrated good potential in plant resilience and increasing metal uptake. Metals in the rhizosphere can be immobilized by AMF, which produce various metabolites that bind HMs, altering soil pH and inhibiting translocation of metals to plant tissues, contributing to both plant health and successful remediation (Ai *et al.*, 2022). Engineered or natural microbial consortia consisting of bacteria and fungi also entail synergistic sequestration, for example fungal effects on bacterial dispersion, increased enzyme activity and co-metabolism of pollutants, resulting in higher degradation efficiencies, particularly those against recalcitrant compounds like polycyclic aromatic hydrocarbons (Cecchi *et al.*, 2021). Considering the circular economy, fungal bioremediation could be improved by actions that aim to harvest metals after remediation and reuse them in industrial processes, thereby changing the remediation process into a value-creating system. Fungal biosorption is effective at capturing HMs, and the reclaimed metals extracted from the fungal biomass after remediation are also very useful. Hence the development of sustainable methodologies to recover metals will be necessary in achieving an economically and environmentally friendly form of fungal remediation (Yarzábal Rodríguez *et al.*, 2024). Studies on extremophilic fungi – fungi that grow in harsh conditions such as low pH, cold, radiation, anaerobe or salty conditions – can be explored for bioremediation potential. Because of antifreeze- and radiation-resistant species such as some Antarctica fungi, even in extreme environments, complex hydrocarbons or even phenolic compounds can be biodegraded. Such characteristics render them useful remediation tools in sites where the use of conventional bioremediation is not possible (Dishliyska *et al.*, 2025).

The other future technology is AI-based biosensor monitoring that may transform real-time identification of the progress of remediation. Biosensors and biosensors facilitated by nanotechnology, although not yet exhaustively used in fungal remediation, are also promising in the detection of pollutants on site, without having to establish laboratories (Banerjee *et al.*, 2021). It is also potentially possible to embed such sensors into fungal bioreactors or polluted sites to monitor changing pollutant levels and fungal activity, leading to a tighter control and optimization of remediation processes.

3.9   Conclusion

Fungi promise an economical and eco-friendly alternative to current remediation processes. This is due to the fact that fungi are extraordinarily adaptive and show remarkable enzymatic versatility. The success of fungi in bioremediation and preserving the ecosystem is achieved through the biosorption, bioaccumulation, enzyme transformation and biomineralization of HMs. Their capability of forming synergetic associations with green plants and bacteria is an added benefit to remediation capability and restoration of ecosystems. On large scales, however, their application suffers setbacks such as inconsistencies in environmental and industrial performance, rates of growth and the costs of recovering metals after adsorption. These limitations can be overcome by developments in immobilization methods, bioreactor methodology optimization and integration into the circular economy model so that resources can be recovered. The future of this area will rely on the usage of emerging technologies to improve fungal strains and via omics-based and nanotechnology-based biosorbents and biosensors powered by AI to provide real-time monitoring. Moreover, fungal extremophiles can be explored to deal with harsher environments. Through further studies, field work testing and conscientious implementation, fungal bioremediation has a lot of potential to improve the health of our global environment through environmental bioremediation.

References

Abdelhameed, R.E. and Metwally, R.A. (2019) Alleviation of cadmium stress by arbuscular mycorrhizal symbiosis. *International Journal of Phytoremediation* 21(7), 663–671. DOI: 10.1080/15226514.2018.1556584.

Ai, Y.-J., Li, F.-P., Yang, J.-Q., Lu, S. and Gu, H.-H. (2022) Research progress and potential functions of AMF and GRSP in the ecological remediation of metal tailings. *Sustainability* 14(15), 9611. DOI: 10.3390/su14159611.

Akhtar, N. and Mannan, M.A.U. (2020) Mycoremediation: Expunging environmental pollutants. *Biotechnology Reports (Amsterdam, Netherlands)* 26, e00452. DOI: 10.1016/j.btre.2020.e00452.

Alengebawy, A., Abdelkhalek, S.T., Qureshi, S.R. and Wang, M.Q. (2021) Heavy metals and pesticides toxicity in agricultural soil and plants: Ecological risks and human health implications. *Toxics* 9(3), 42. DOI: 10.3390/toxics9030042.

Ali, E.H. and Hashem, M. (2007) Removal efficiency of the heavy metals Zn(II), Pb(II) and Cd(II) by *Saprolegnia delica* and *Trichoderma viride* at different pH values and temperature degrees. *Mycobiology* 35(3), 135–144. DOI: 10.4489/MYCO.2007.35.3.135.

Ali, E.A.M., Sayed, M.A., Abdel-Rahman, T.M.A. and Hussein, R. (2021) Fungal remediation of Cd(ii) from wastewater using immobilization techniques. *RSC Advances* 11(8), 4853–4863. DOI: 10.1039/d0ra08578b.

Alshiekheid, M.A., Umar, A., Ameen, F., Alyahya, S.A. and Dufossé, L. (2023) Biodegradation of chromium by laccase action of *Ganoderma multipileum*. *Journal of King Saud University – Science* 35(10), 102948. DOI: 10.1016/j.jksus.2023.102948.

Amobonye, A., Aruwa, C.E., Aransiola, S., Omame, J., Alabi, T.D. *et al.* (2023) The potential of fungi in the bioremediation of pharmaceutically active compounds: A comprehensive review. *Frontiers in Microbiology* 14, 1207792. DOI: 10.3389/fmicb.2023.1207792.

Angeles de Paz, G., Martínez-Gutierrez, H., Ramírez-Granillo, A., López-Villegas, E.O., Medina-Canales, M.G. *et al.* (2023) *Rhodotorula mucilaginosa* YR29 is able to accumulate Pb^{2+} in vacuoles: A yeast with bioremediation potential. *World Journal of Microbiology and Biotechnology* 39(9), 238. DOI: 10.1007/s11274-023-03675-4.

Atila, F. and Kazankaya, A. (2023) Evaluation of the yield and heavy metal bioaccumulation in the fruit body of *Pleurotus ostreatus* grown on sugar mill wastewaters. *Biomass Conversion and Biorefinery* 14(16), 19177–19186. DOI: 10.1007/s13399-023-03913-7.

Ayele, A., Haile, S., Alemu, D. and Kamaraj, M. (2021) Comparative utilization of dead and live fungal biomass for the removal of heavy metal: A concise review. *Scientific World Journal* 5588111. DOI: 10.1155/2021/5588111.

Banerjee, A., Maity, S. and Mastrangelo, C.H. (2021) Nanotechnology for biosensors: A review. *arXiv Preprint arXiv*. DOI: 10.20944/preprints202101.0152.v1.

Beltrán-Flores, E., Pla-Ferriol, M., Martínez-Alonso, M., Gaju, N., Sarrà, M. *et al.* (2023) Fungal treatment of agricultural washing wastewater: Comparison between two operational strategies. *Journal of Environmental Management* 325(Pt A), 116595. DOI: 10.1016/j.jenvman.2022.116595.

Bhardwaj, A. (2025) Endophytes unleashed: Natural allies for heavy metal bioremediation. *Discover Plants* 2(1), 57. DOI: 10.1007/s44372-025-00137-z.

Bhattacharya, A., Gola, D., Dey, P. and Malik, A. (2020) Synergistic and antagonistic effects on metal bioremediation with increasing metal complexity in a hexa-metal environment by *Aspergillus fumigatus*. *International Journal of Environmental Research* 14(6), 761–770. DOI: 10.1007/s41742-020-00295-w.

Briffa, J., Sinagra, E. and Blundell, R. (2020) Heavy metal pollution in the environment and their toxicological effects on humans. *Heliyon* 6(9), e04691.

Buechel, E.R. and Pinkett, H.W. (2020) Transcription factors and ABC transporters: From pleiotropic drug resistance to cellular signaling in yeast. *FEBS Letters* 594(23), 3943–3964. DOI: 10.1002/1873-3468.13964.

Cai, C.X., Xu, J., Deng, N.F., Dong, X.W., Tang, H. *et al.* (2016) A novel approach of utilization of the fungal conidia biomass to remove heavy metals from the aqueous solution through immobilization. *Scientific Reports* 6, 36546. DOI: 10.1038/srep36546.

Cavalheiro, M., Pais, P., Galocha, M. and Teixeira, M.C. (2018) Host-pathogen interactions mediated by MDR transporters in fungi: As pleiotropic as it Gets! *Genes* 9(7), 332. DOI: 10.3390/genes9070332.

Cecchi, G., Piazza, S., Rosatto, S., Mariotti, M.G., Roccotiello, E. *et al.* (2021) A mini-review on the co-growth and interactions among microorganisms (fungi and bacteria) from rhizosphere of metal-hyperaccumulators. *Frontiers in Fungal Biology* 2, 787381. DOI: 10.3389/ffunb.2021.787381.

Chen, L., Zhang, X., Zhang, M., Zhu, Y. and Zhou, R. (2022) Removal of heavy-metal pollutants by white-rot fungi: Recent achievements. *Journal of Cleaner Production* 354, 131681. DOI: 10.1016/j.jclepro.2022.131681.

Chiapello, M., Martino, E. and Perotto, S. (2015) Common and metal-specific proteomic responses to cadmium and zinc in the metal tolerant ericoid mycorrhizal fungus *Oidiodendron maius* Zn. *Metallomics* 7(5), 805–815. DOI: 10.1039/c5mt00024f.

Corbu, V.M., Gheorghe-Barbu, I., Dumbravă, A.Ş., Vrâncianu, C.O. and Şesan, T.E. (2023) Current insights in fungal importance-A comprehensive review. *Microorganisms* 11(6), 1384. DOI: 10.3390/microorganisms11061384.

Dagdag, O., Quadri, T.W., Haldhar, R., Kim, S.C., Daoudi, W. *et al.* (2023) An overview of heavy metal pollution and control. In: *Heavy Metals in the Environment: Management Strategies for Global Pollution*. American Chemical Society, pp. 3–24.

Dinakarkumar, Y., Ramakrishnan, G., Gujjula, K.R., Vasu, V., Balamurugan, P. *et al.* (2024) Fungal bioremediation: An overview of the mechanisms, applications and future perspectives. *Environmental Chemistry and Ecotoxicology* 6, 293–302. DOI: 10.1016/j.enceco.2024.07.002.

Dishliyska, V., Miteva-Staleva, J., Gocheva, Y., Stoyancheva, G., Yovchevska, L. *et al.* (2025) Biological potential of extremophilic filamentous fungi for the production of new compounds with antimicrobial effect. *Fermentation* 11(6), 347. DOI: 10.3390/fermentation11060347.

Dusengemungu, L., Kasali, G., Gwanama, C. and Ouma, K.O. (2020) Recent advances in biosorption of copper and cobalt by filamentous fungi. *Frontiers in Microbiology* 11, 582016. DOI: 10.3389/fmicb.2020.582016.

Elahian, F., Heidari, R., Charghan, V.R., Asadbeik, E. and Mirzaei, S.A. (2020) Genetically modified *Pichia pastoris*, a powerful resistant factory for gold and palladium bioleaching and nanostructure heavy metal biosynthesis. *Artificial Cells, Nanomedicine, and Biotechnology* 48(1), 259–265. DOI: 10.1080/21691401.2019.1699832.

El-Gendi, H., Saleh, A.K., Badierah, R., Redwan, E.M., El-Maradny, Y.A. *et al.* (2021) A comprehensive insight into fungal enzymes: Structure, classification, and their role in mankind's challenges. *Journal of Fungi (Basel, Switzerland)* 8(1), 23. DOI: 10.3390/jof8010023.

El-Gendy, M.M.A.A., Abdel-Moniem, S.M., Ammar, N.S. and El-Bondkly, A.M.A. (2023) Bioremoval of heavy metals from aqueous solution using dead biomass of indigenous fungi derived from fertilizer industry effluents: Isotherm models evaluation and batch optimization. *Biometals* 36(6), 1307–1329. DOI: 10.1007/s10534-023-00520-x.

El-Sharkawy, M., Alotaibi, M.O., Li, J., Du, D. and Mahmoud, E. (2025) Heavy metal pollution in coastal environments: Ecological implications and management strategies: A review. *Sustainability* 17(2), 701. DOI: 10.3390/su17020701.

Feng, B., Xue, Y., Wang, D., Chen, S., Zhang, S. *et al.* (2025) Stability of lead immobilization by *Aspergillus niger* and fluorapatite under different pH conditions. *Ecotoxicology and Environmental Safety* 289, 117706. DOI: 10.1016/j.ecoenv.2025.117706.

Fouda-Mbanga, B.G., Onotu, O.P. and Tywabi-Ngeva, Z. (2024) Advantages of the reuse of spent adsorbents and potential applications in environmental remediation: A review. *Green Analytical Chemistry* 11, 100156. DOI: 10.1016/j.greeac.2024.100156.

He, M., Xu, Y., Qiao, Y., Zhang, Z., Liang, J. *et al.* (2022) A novel yeast strain *Geotrichum* sp. CS-67 capable of accumulating heavy metal ions. *Ecotoxicology and Environmental Safety* 236, 113497. DOI: 10.1016/j.ecoenv.2022.113497.

Hu, Y., Wang, D. and Li, Y. (2016) Environmental behaviors and potential ecological risks of heavy metals (Cd, Cr, Cu, Pb, and Zn) in multimedia in an oilfield in China. *Environmental Science and Pollution Research* 23(14), 13964–13972. DOI: 10.1007/s11356-016-6589-1.

Iram, S., Parveen, K., Usman, J., Nasir, K., Akhtar, N. *et al.* (2012) Heavy metal tolerance of filamentous fungal strains isolated from soil irrigated with industrial wastewater. *Biologija* 58(3), 107–116.

Kaur, P., Sharma, S., Albarakaty, F.M., Kalia, A., Hassan, M.M. *et al.* (2022) Biosorption and bioleaching of heavy metals from electronic waste varied with microbial genera. *Sustainability* 14(2), 935. DOI: 10.3390/su14020935.

Khandelwal, N.K., Millan, C.R., Zangari, S.I., Avila, S., Williams, D. *et al.* (2022) The structural basis for regulation of the glutathione transporter Ycf1 by regulatory domain phosphorylation. *Nature Communications* 13(1), 1278. DOI: 10.1038/s41467-022-28811-w.

Kondakindi, V.R., Pabbati, R., Erukulla, P., Maddela, N.R. and Prasad, R. (2024) Bioremediation of heavy metals-contaminated sites by microbial extracellular polymeric substances – A critical view. *Environmental Chemistry and Ecotoxicology* 6, 408–421. DOI: 10.1016/j.enceco.2024.05.002.

Kumar, M., Seth, A., Singh, A.K., Rajput, M.S. and Sikandar, M. (2021) Remediation strategies for heavy metals contaminated ecosystem: A review. *Environmental and Sustainability Indicators* 12, 100155. DOI: 10.1016/j.indic.2021.100155.

Kuppan, N., Padman, M., Mahadeva, M., Srinivasan, S. and Devarajan, R. (2024) A comprehensive review of sustainable bioremediation techniques: Eco friendly solutions for waste and pollution management. *Waste Management Bulletin* 2(3), 154–171. DOI: 10.1016/j.wmb.2024.07.005.

Lanfranco, L. and Young, J.P.W. (2012) Genetic and genomic glimpses of the elusive arbuscular mycorrhizal fungi. *Current Opinion in Plant Biology* 15(4), 454–461. DOI: 10.1016/j.pbi.2012.04.003.

Laoye, B., Olagbemide, P., Ogunnusi, T. and Akpor, O. (2025) Heavy metal contamination: Sources, health impacts, and sustainable mitigation strategies with insights from Nigerian case studies. *F1000Research* 14, 134. DOI: 10.12688/f1000research.160148.4.

Li, Q., Zhang, M., Wei, B., Lan, W., Wang, Q. *et al.* (2024) Fungal biomineralization of toxic metals accelerates organic pollutant removal. *Current Biology* 34(10), 2077–2084. DOI: 10.1016/j.cub.2024.04.005.

Li, Z.S., Lu, Y.P., Zhen, R.G., Szczypka, M., Thiele, D.J. *et al.* (1997) A new pathway for vacuolar cadmium sequestration in *Saccharomyces cerevisiae*: YCF1-catalyzed transport of bis (glutathionato) cadmium. *Proceedings of the National Academy of Sciences* 94(1), 42–47.

Liu, Q., Bai, J., Li, R., Gu, W., Peng, S. *et al.* (2022) Electrochemical oxidation of copper-clad laminate for manufacturing printed circuit boards via bioleaching by the fungus *Phanerochaete chrysosporium* 144, 108002. DOI: 10.1016/j.bioelechem.2021.108002.

MacDiarmid, C.W., Gaither, L.A. and Eide, D. (2000) Zinc transporters that regulate vacuolar zinc storage in *Saccharomyces cerevisiae*. *EMBO Journal* 19(12), 2845–2855. DOI: 10.1093/emboj/19.12.2845.

Maglione, G., Zinno, P., Tropea, A., Mussagy, C.U., Dufossé, L. *et al.* (2024) Microbes' role in environmental pollution and remediation: A bioeconomy focus approach. *AIMS Microbiology* 10(3), 723–755. DOI: 10.3934/microbiol.2024033.

Mohamadhasani, F. and Rahimi, M. (2022) Growth response and mycoremediation of heavy metals by fungus *Pleurotus* sp. *Scientific Reports* 12(1), 19947. DOI: 10.1038/s41598-022-24349-5.

Mousa, A.M., Breky, M.M.E. and Attallah, M.F. (2025) Biosorption of cesium and strontium from aqueous solution by *Aspergillus flavus* biomass. *Scientific Reports* 15(1), 26328. DOI: 10.1038/s41598-025-11603-9.

Mulligan, C.N. and Galvez-Cloutier, R. (2003) Bioremediation of metal contamination. *Environmental Monitoring and Assessment* 84(1–2), 45–60. DOI: 10.1023/a:1022874727526.

Pande, V., Pandey, S.C., Sati, D., Bhatt, P. and Samant, M. (2022) Microbial interventions in bioremediation of heavy metal contaminants in agroecosystem. *Frontiers in Microbiology* 13, 824084.

Parasnis, M.S., Deng, E., Yuan, M., Lin, H., Kordas, K. *et al.* (2024) Heavy metal remediation by dry mycelium membranes: Approaches to sustainable lead remediation in water. *Langmuir* 40(12), 6317–6329. DOI: 10.1021/acs.langmuir.3c03811.

Paria, K., Pyne, S. and Chakraborty, S.K. (2022) Optimization of heavy metal (lead) remedial activities of fungi *Aspergillus penicillioides* (F12) through extra cellular polymeric substances. *Chemosphere* 286(Pt 3), 131874. DOI: 10.1016/j. chemosphere.2021.131874.

Parvin, S., Geel, M., Yeasmin, T., Lievens, B. and Honnay, O. (2019) Variation in arbuscular mycorrhizal fungal communities associated with lowland rice (*Oryza sativa*) along a gradient of soil salinity and arsenic contamination in Bangladesh. *Science of the Total Environment* 686, 546–554. DOI: 10.1016/j. scitotenv.2019.05.450.

Ramya, D., Kiruba, N.J.M. and Thatheyus, A.J. (2021) Biosorption of heavy metals using fungal biosorbents – A review. In: *Fungi Bio-Prospects in Sustainable Agriculture, Environment and Nano-Technology*. pp. 331–352. DOI: 10.1016/ B978-0-12-821925-6.00015-0.

Rezghi Rami, M., Meskini, M. and Ebadi Sharafabad, B. (2024) Fungal-mediated nano-particles for industrial applications: Synthesis and mechanism of action. *Journal of Infection and Public Health* 17(10), 102536. DOI: 10.1016/j.jiph.2024.102536.

Rostami, K.H. and Joodaki, M.R. (2002) Some studies of cadmium adsorption using *Aspergillus niger, Penicillium austurianum*, employing an airlift fermenter. *Chemical Engineering Journal* 89(1–3), 239–252. DOI: 10.1016/S1385-8947(02)00131-6.

Roy, A., Gogoi, N., Haider, F.U. and Farooq, M. (2025) Mycoremediation for sustain-able remediation of environmental pollutants. *Biocatalysis and Agricultural Biotechnology* 64, 103526. DOI: 10.1016/j.bcab.2025.103526.

Rudakiya, D.M., Iyer, V., Shah, D., Gupte, A. and Nath, K. (2018) Biosorption potential of *Phanerochaete chrysosporium* for arsenic, cadmium, and chromium removal from aqueous solutions. *Global Challenges* 2(12), 1800064. DOI: 10.1002/ gch2.201800064.

Sánchez-Castellón, J., Urango-Cárdenas, I., Enamorado-Montes, G., Burgos-Nuñez, S., Marrugo-Negrete, J. *et al.* (2022) Removal of mercury, cadmium, and lead ions by *Penicillium* sp. *Frontiers in Environmental Chemistry* 2, 795632. DOI: 10.3389/ fenvc.2021.795632.

Seh-Bardan, B.J., Othman, R., Ab Wahid, S., Husin, A. and Sadegh-Zadeh, F. (2012) Column bioleaching of arsenic and heavy metals from gold mine tailings by *Aspergillus fumigatus. CLEAN – Soil, Air, Water* 40(6), 607–614. DOI: 10.1002/ clen.201000604.

Shao, Q., Yan, S., Sun, X., Chen, H., Lu, Y. *et al.* (2025) Applications of yeasts in heavy metal remediation. *Fermentation* 11(5), 236. DOI: 10.3390/ fermentation11050236.

Sharma, K.R., Giri, R. and Sharma, R.K. (2023) Efficient bioremediation of metal containing industrial wastewater using white rot fungi. *International Journal of Environmental Science and Technology* 20(1), 943–950.

Sidhu, A.K., Agrawal, S.B., Verma, N., Kaushal, P. and Sharma, M. (2025) Fungal-mediated synthesis of multimetallic nanoparticles: Mechanisms, unique proper-ties, and potential applications. *Frontiers in Nanotechnology* 7, 1549713. DOI: 10.3389/fnano.2025.1549713.

Singh, A., Sharma, R.K., Agrawal, M. and Marshall, F.M. (2010) Health risk assessment of heavy metals via dietary intake of foodstuffs from the wastewater irrigated site of a dry tropical area of India. *Food and Chemical Toxicology* 48(2), 611–619. DOI: 10.1016/j.fct.2009.11.041.

Singh, V.K. and Singh, R. (2024) Role of white rot fungi in sustainable remediation of heavy metals from the contaminated environment. *Mycology* 15(4), 585–601. DOI: 10.1080/21501203.2024.2389290.

Siracusa, G., Becarelli, S., Lorenzi, R., Gentini, A. and Gregorio, S. (2017) PCB in the environment: Bio-based processes for soil decontamination and management of waste from the industrial production of *Pleurotus ostreatus*. *New Biotechnology* 39(Pt B), 232–239. DOI: 10.1016/j.nbt.2017.08.011.

Stella, T., Covino, S., Čvančarová, M., Filipová, A., Petruccioli, M. *et al.* (2017) Bioremediation of long-term PCB-contaminated soil by white-rot fungi. *Journal of Hazardous Materials* 324(Pt B), 701–710. DOI: 10.1016/j.jhazmat.2016.11.044.

Thepbandit, W. and Athinuwat, D. (2024) Rhizosphere microorganisms supply availability of soil nutrients and induce plant defense. *Microorganisms* 12(3), 558. DOI: 10.3390/microorganisms12030558.

Tian, Q., Wang, J., Shao, S., Zhou, H., Kang, J. *et al.* (2024) Combining metabolomics and transcriptomics to analyze key response metabolites and molecular mechanisms of *Aspergillus fumigatus* under cadmium stress. *Environmental Pollution* 356, 124344. DOI: 10.1016/j.envpol.2024.124344.

Vig, I., Benkő, Z., Gila, B.C., Palczert, Z., Jakab, Á. *et al.* (2023) Functional characterization of genes encoding cadmium pumping P1B-type atpases in *Aspergillus fumigatus* and *Aspergillus nidulans*. *Microbiology Spectrum* 11(5), e00283–23. DOI: 10.1128/spectrum.00283-23.

Yan, G. and Viraraghavan, T. (2003) Heavy-metal removal from aqueous solution by fungus *Mucor rouxii*. *Water Research* 37(18), 4486–4496. DOI: 10.1016/S0043-1354(03)00409-3.

Yang, H.Y., Liu, Q., Chen, G.B., Tong, L.L. and Ali, A. (2018) Bio-dissolution of pyrite by *Phanerochaete chrysosporium*. *Transactions of Nonferrous Metals Society China* 28(4), 766–774.

Yang, Y., Liu, R., Zhou, Y., Tang, Y., Zhang, J. *et al.* (2024) Screening and performance optimization of fungi for heavy metal adsorption in electrolytes. *Frontiers in Microbiology* 15, 1371877. DOI: 10.3389/fmicb.2024.1371877.

Yarzábal Rodríguez, L.A., Álvarez Gutiérrez, P.E., Gunde-Cimerman, N., Ciancas Jiménez, J.C., Gutiérrez-Cepeda, A. *et al.* (2024) Exploring extremophilic fungi in soil mycobiome for sustainable agriculture amid global change. *Nature Communications* 15(1), 6951. DOI: 10.1038/s41467-024-51223-x.

Zareh, M.M., El-Sayed, A.S. and El-Hady, D.M. (2022) Biosorption removal of iron from water by *Aspergillus niger*. *NPJ Clean Water* 5(1), 58. DOI: 10.1038/s41545-022-00201-1.

Zhang, X., Hu, W., Xie, X., Wu, Y., Liang, F. *et al.* (2021) Arbuscular mycorrhizal fungi promote lead immobilization by increasing the polysaccharide content within pectin and inducing cell wall peroxidase activity. *Chemosphere* 267, 128924. DOI: 10.1016/j.chemosphere.2020.128924.

Zhang, X., Liu, Y., Mo, X., Huang, Z., Zhu, Y. *et al.* (2025) Ectomycorrhizal fungi and biochar promote soil recalcitrant carbon increases under arsenic stress. *Journal of Hazardous Materials* 489, 137598. DOI: 10.1016/j.jhazmat.2025.137598.

Zhao, S., Yan, L., Kamran, M., Liu, S. and Riaz, M. (2024) Arbuscular mycorrhizal fungi-assisted phytoremediation: A promising strategy for cadmium-contaminated soils. *Plants* 13(23), 3289. DOI: 10.3390/plants13233289.

Algae-based Methods for Remediation of Heavy Metals from Polluted Sites: Recent Updates

4

4.1 Introduction

Contamination of heavy metals (HMs) in aquatic ecosystems is a major concern for the environment. Metals like lead (Pb), cadmium (Cd), copper (Cu), zinc (Zn), mercury (Hg), chromium (Cr), cobalt (Co) and arsenic (As) are widely present in the environment and cause adverse impacts (Masindi and Muedi, 2018). Nowadays both organic contaminants and HMs are regarded as major environmental concerns as their pollution poses adverse impacts on ecosystems and sustainability. These pollutants have originated from different types of industrial activities, the improper release of domestic wastewater and from other wastes. Anthropogenic activity has led to increased pollution of soil and aquatic systems with harmful metals and hazardous substances (Bala *et al.*, 2022; Tripathi *et al.*, 2025). HMs are contaminating marine waters and adversely affecting life forms in marine ecosystems (Tripathi *et al.*, 2025). Therefore, it is essential to reduce the HM burden from contaminated environments before discharging them into other water channels (Tripathi *et al.*, 2011; Garg *et al.*, 2012).

Multiple clean-up strategies including traditional ones like ion exchange, precipitation, chemical extraction, etc., have been used to remove HMs. However, these methods are costly for large-scale applications, and have less HM remediation efficiency. Algae-based bioremediation of heavy metals is a sustainable, environmentally friendly and feasible option of pollutant clean-up (Salama *et al.*, 2019; Znad *et al.*, 2022; Parmar and Patel, 2025). Different types of algae including micro- and macroalgae are used in bioremediation of HMs using multiple mechanisms like biosorption, bioaccumulation, complexation, ion exchange, precipitation, etc. Among these methods, biosorption may be the best strategy because it is cost-effective, highly efficient and easy to operate (Torres, 2020). However, biotreatment with microalgae is also gaining popularity among researchers seeking sustainable contaminated water treatment

Corresponding author: manikant.microbio@gmail.com

methods (Utomo *et al.*, 2016). A recent review by Salimi *et al.* (2025) discussed remediation of HM contaminants by biosorption and bioaccumulation techniques using the algae *Chlorella* sp. and *Sargassum* sp. Different environmental factors like temperature, pH, biomass dose and other cultural conditions affect the bioremediation efficiency of algae for HM remediation from polluted sites.

This chapter provides an overview of the application of algae for biological remediation of toxic HMs along with the mechanisms involved in phycoremediation.

Table 4.1 shows the different effects of heavy metals on the human body.

4.2 Phycoremediation of Heavy Metals

4.2.1 Heavy metal removal from contaminated environments

Algae play an important role in eliminating HMs from contaminated environments (Salama *et al.*, 2019). Algae may utilize the waste materials present at polluted sites as a nutrient source, helping reduce different types of pollution through various enzymatic and metabolic processes. Salama *et al.* (2019) studied both living and non-living algal cells and found them to be capable of binding HMs from contaminated environments. They do this through several methods such as adsorption, complexation, chemisorption, chelation and reduction. Algal metabolic pathways play a crucial role in managing the pollution by breaking down, detoxifying and transforming HMs and harmful

Table 4.1. Different health effects due to heavy metal toxicity.

Heavy metals	Health effects	Reference
Mercury (Hg)	Reproductive disorders, skin lesions, vision damage, neurological disorders	Balali-Mood *et al.*, 2021; Charkiewicz *et al.*, 2025
Lead (Pb)	Decreased liver function, changes in excretory function, brain damage, circulatory disorders	Kumar *et al.*, 2020; Tizabi *et al.*, 2023
Arsenic (As)	Associated with chronic diseases such as cancer, cardiovascular disorders and neurotoxicity	Vellingiri *et al.*, 2022; Ganie *et al.*, 2023
Cadmium (Cd)	Kidney failure, bone lesions	Ding *et al.*, 2023
Chromium (Cr)	Dermal toxicity, respiratory troubles, lung cancer	Alvarez *et al.*, 2021; Hagvall *et al.*, 2021; Haidar *et al.*, 2023
Nickel (Ni)	Dermatitis, chronic asthma	Koh *et al.*, 2019
Zinc (Zn)	Gastrointestinal disorders, nausea, loss of appetite, lethargy	Aralbaeva *et al.*, 2024
Manganese (Mn)	Neurotoxicity	Dey *et al.*, 2023

synthetic compounds (xenobiotics), sometimes even converting them into less harmful or volatile forms (Ankit and Korstad, 2022). Biosorption is a passive process in which adsorption of metal occurs (Garg *et al.*, 2012; Ahmad *et al.*, 2020). Algae are autotrophic in nature, have high surface area to volume ratios and have genetic manipulation potential and phytochelatin expression (Saini and Dhania, 2020). Various algae are reported as being useful for the bioremediation of toxic HMs in contaminated environment (Table 4.2).

4.2.2   Mechanisms of phycoremediation

Heavy metal remediation through algae can be carried out through bioremediation techniques. Microalgae also use biosorption and

Table 4.2. Bioremoval of heavy metals by different algae.

Algae used	Heavy metal remediated	Reference
Chlorella vulgaris	Iron (Fe^{2+}), manganese (Mn^{2+}) and zinc (Zn^{2+})	Salimi *et al.*, 2025
Pretreated biomass of red algae *Digenia simplex*	Remediation of 97.27% cadmium (Cd^{2+}) at pH 5.78, initial metal 24.79 mg/l concentration and adsorbent dosage 6.13 g/l	Hassan *et al.*, 2025
Microcystis aeruginosa	83.24% remediation of copper (Cu) ions at initial concentration 25 mg/l, 30°C, pH 8 and adsorption time 5 h	Zeng *et al.*, 2022
Red algae *Porphyra leucosticta* biomass	70% remediation for Cd(II) and 90% for Pb(II)	Ye *et al.*, 2015
Microalgae biomass with ferric oxide magnetic nanoparticles	58% remediation for Cr(VI) and 73.4% for Cu(II) at pH 6, initial concentration 5 mg/l Cr(VI) and 20 mg/l Cu(II) and contact time 6 h	Lin *et al.*, 2024
Cladophora glomerata, *Oedogonium westii*, *Vaucheria debaryana* and *Zygnema insigne*	Removal of chromium (Cr), Cd and lead (Pb) through bioaccumulation process	Shamshad *et al.*, 2015
Chlorella sp.	Remediation of 67% Cu, 50% Cr, 69% Pb and 93% Cd	Kumar *et al.*, 2013
Oedogonium westi	Removal efficiencies for Cd, Cr, nickel (Ni) and Pb were 55–95%, 61–93%, 59–89% and 61–96%, respectively	Shamshad *et al.*, 2015
Galdieria sulphuraria	30% remediation in medium containing 3 mg/l of Cd ions	Kharel *et al.*, 2024
Spirogyra hyalina dried biomass	Remediation of Cd, mercury (Hg), Pb, arsenic (As) and Co	Kumar and Oommen, 2012

bioaccumulation processes to remove HM ions. Fig. 4.1 shows the biosorption and bioaccumulation methods used in the treatment of HMs.

Many researchers have studied and observed phycoremediation of HMs. Khan *et al.* (2025) tested whether an algal net biomass consisting of either a natural mixed algal community or a pure microalgal strain (*Asterarcys* sp. RA100) could bioremediate Cr(VI) in contaminated water. They observed the effect of pH, temperature and contact time and found that both algal biomasses are extremely efficient, cost-effective and eco-friendly methods when compared with chemical treatments for maximal bioremediation of Cr(VI)-polluted water and also has a great potential for industrial scale-up. Another study, by Ciempiel *et al.* (2025), observed that the extracellular polymeric substances (EPS) obtained from mixotrophically grown algae *Chlorella vulgaris*, *Parachlorella kessleri* and *Vischeria magna* efficiently remediated Pb(II) from aqueous solution. Of these algae, EPS from *C. vulgaris* demonstrated the maximum sorption capacity of Pb(II), highlighting its effectiveness in the removal of Pb from contaminated environments. Similarly, Tenza *et al.* (2025), studied the bioremediation of heavy metals such as Cu, Pb and Zn from wastewater by *Chlorella* sp. biomass. Researchers observed that the optimal removal of HMs was obtained at pH 7 with a contact time of 60 min, attaining full or almost full HM remediation for wastewater containing HMs. The adsorption of Cu, Pb and Zn was performed by various functional groups, for instance hydroxyl, amide, carboxyl and carbonyl, present on the surface of the algal biomass. These results suggest that the *Chlorella* sp. algal biomass is a resilient

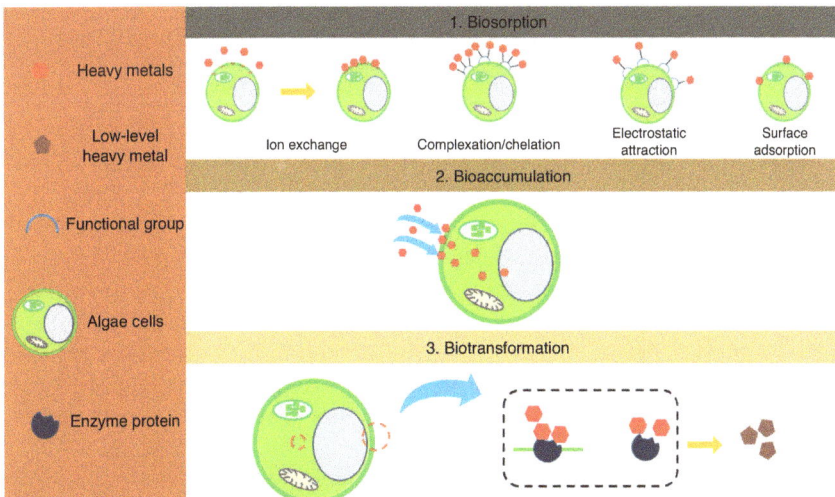

Fig. 4.1. Mechanisms of phycoremediation of heavy metals using various methods. From Chen *et al.* (2023).

and competent tool for the bioremediation of HMs, but still requires more in-depth research on the scaling up of its wastewater treatment application.

Research by Liu *et al.* (2024) also demonstrated the high tolerance of hal-otolerant microalga *Dunaliella* sp. FACHB-558 to Co, nickel (Ni) and Cd. This study also showed effective absorption of Co ions from aquatic environments suggesting that this microalga could be an eco-conscious, green biosorbent for the eco-friendly remediation of HMs in saline and coastal waters, with the additional benefit of carbon sequestration and energy efficiency. In a recent review, Tripathi *et al.* (2025) discussed the role of microorganisms in bio-based remediation of various heavy metals toxicants from marine ecosystem that helps in environmental sustainability. *Algae based heavy metals* removal is one of the ecofriendly approaches. While environmental factors affect the heavy metals phycoremediation efficiency, advanced strategies like genetic engineering might be helpful in the enhanced bioremediation performance. Singh *et al.* (2024) observed a similar result with acclimatized microalgae *Arthrospira platensis* and *Spirulina* sp., which efficiently removed HMs such as Cr and Cd and assisted in the biodegradation of dyes such as methyl orange and crystal violet from synthetic wastewater. It was reported that the removal of HMs was found to be 100% within 24 h, highlighting the potential of algal bioremediation as a sustainable and green strategy for the treatment of industrial wastewater.

In another study Ganguly *et al.* (2024) observed that an isolated strain of microalgae *Chlorella thermophilia* that was collected, identified and cultured from a natural source. It was investigated as a bioremediating agent from the removal of Cr(VI) and As(III) from wastewater; 50–65% removal of Cr(VI) and 70–91% removal of As(III) was observed. A shift in the biochemical composition was also observed, suggesting the generation of carbohydrate-rich biomass. *C. thermophilia* can also therefore be considered as a potential candidate for effective bioremediation of toxic HMs. Oyebamiji *et al.* (2021) conducted a study utilizing locally isolated microalgal strains from Nigeria, including *Chlorella* sp. MOW12, to study the tolerance and bioremoval of Pb from aqueous solutions. It was observed that all the four Nigerian strains survived Pb concentrations of up to 60 ppm. The level of Pb removal was found to be high, extending from 86% to 93%, with *Chlorella* showcasing the highest capacity of Pb removal. Overall, the research confirmed that area-specific microalgae are promising candidates for the efficient and sustainable bioremediation of HMs. Other researchers, Diaconu *et al.* (2023), studied and observed that the macroalgae *Sargassum fusiforme* and *Enteromorpha prolifera* were effective in the removal of HMs particularly Pb and manganese (Mn). Removal of 100% of Pb and 98.17% of Mn was achieved by *E. prolifera* over a period of 10 h. *S. fusiforme* achieved 99.46% removal of Pb and 95.73% of Mn.

Sbihi *et al.* (2024) studied the bioremediation of Cr(VI) from wastewater through a freshwater microalga *Craticula subminuscula* obtained from a Moroccan river located in the High Atlas mountains. The microalga illustrated

95.32% removal of Cr(VI) under optimal conditions of pH 1, adsorbent dose of 10.91 mg/l and time of 129.47 min. Fourier transformer infrared spectroscopy (FTIR) analysis of the biosorbent indicated the presence of functional groups such as amides, carboxylic acids, aldehydes, phosphates and halides, which assist in the binding of Cr(VI) from wastewater onto the surface of the biosorbent. This study also demonstrated the viability of repeated cycles of desorption, therefore encouraging the design of scalable, continuous biosorption processes.

Therefore, algae-based treatment can be a promising, viable, budget-friendly, green and effective strategy for removing HMs with possible futuristic potential of on-site treatment of polluted aquatic environments.

Biosorption is defined as the process of removing chemicals from a solution using various biological materials to reduce the concentration of sorbate in the solution and to accumulate the sorbate at the sorbate–sorbent interface (Luka *et al.*, 2024). Biosorption is a physiochemical phenomenon that results in the elimination of HMs by covalent or ionic interactions. Metal ion biosorption is facilitated by involvement of different chemical functional groups. Algal cell walls contain a net negative charge due to the presence of functional groups like PO_4^{3-} and $COO-$ that interact with HMs via ion exchange (Spain *et al.*, 2021). The pH of the adsorbing media has a substantial influence on metal ion adsorption. Alkaline pH increases the attraction of metal cations and, eventually, their adsorption on cell surfaces with a negative charge, such as phosphate, polysaccharides, aminos, carboxyl groups of protein and amino groups of nucleic acid (Zhao *et al.*, 2023). Bioaccumulation is a dependent metabolic mechanism that refers to the accumulation of HMs within the cell membranes of living microalgae with active and passive transport channels (Shamim, 2018). Researchers have investigated the use of immobilized algae for remediation of HMs from polluted water (Chen *et al.*, 2023). Both living and non-living algal biomasses effectively remediate HM toxicants from polluted aquatic environments (Ahmad *et al.*, 2020; Greeshma *et al.*, 2022). In another study, Barquilha *et al.* (2017) applied free algae and immobilized algae for biosorption of metal (Pb and Cu) ions. These studies indicate the possibility of using algae in remediation of HMs from industrial wastewater or contaminated environments.

Shamshad *et al.* (2015) studied bioaccumulation of HMs from aquatic environments by using freshwater algae like *Cladophora glomerata*, *Oedogonium westii*, *Vaucheria debaryana* and *Zygnema insigne*. They assessed the bioaccumulation capability of these algae for HMs like Cd, Cr and Pb. Researchers reported algae-based biosorption of toxic HMs followed by intracellular accumulation through metal transporters, and enzymatic biotransformation (Danouche *et al.*, 2021). In another study, Znad *et al.* (2022) discussed the use of algae and seaweed biomass for remediating HMs from wastewater, and suggested that modifying the preparation methods of biomass before the adsorption process may enhance bioremediation efficiency for possible large-scale wastewater treatment using algal biomass. Besides microbial applications for decontamination, waste biomass may also be use in the treatment process. In a recent review, Tripathi *et al.* (2024) discussed strategies for remediation of hexavalent Cr by using adsorbents like waste biomass. Other researchers have explored the use of microbial biomass for the abatement of various types of inorganic and organic pollutants from affected environmental sites. Operational parameters require monitoring for optimal bioremediation performance. Integration with other advanced technologies like nanotechnology may be helpful in enhancing remediation of such toxicants from aquatic environment.

4.3   Challenges and Future Prospects

Algae-based biotreatment is a promising, environmentally friendly approach. The process parameters affect phycoremediation efficiency, so the impact of environmental factors need to be monitored carefully and optimized for effective bioremoval of HMs from contaminated sites. Mechanistic understanding of how algae or algal biomass is involved in remediation of HMs is also very important for effective bioremediation. *In situ* large-scale application of algae-based remediation of specific HMs may face challenges due to the presence of other coexisting inorganic and organic pollutants in real wastewater. Integration with other advanced methods or by using consortia of microbes might help overcome these issues. More research is required for molecular and genetic understanding of algae-based treatment. Genetic modification can improve the strain's capability, ultimately resulting in enhanced bioremediation efficiency. Consortia of algae with other microbes like bacteria may also be beneficial for HM removal from contaminated sites.

4.4   Conclusion

Toxicity of both inorganic and organic pollutants are environmental threats that adversely affect plants, animals and humans. HMs are a major threats to ecosystems. These toxic chemicals can be remediated by algae through biosorption, bioaccumulation and enzymatic detoxification. Various operational environmental parameters affect the efficiency of algae-based remediation of toxic metals from contaminated sites. The rate of HM detoxification may be enhanced by using improved algae through genetic engineering or bioengineering the microorganisms. More research attention is required examining remediation of environmental toxicants via integrated treatment strategies with algae in an environmentally friendly manner.

References

Ahmad, S., Pandey, A., Pathak, V.V., Tyagi, V.V. and Kothari, R. (2020) Phycoremediation: Algae as eco-friendly tools for the removal of heavy metals from wastewaters. In: *Bioremediation of Industrial Waste for Environmental Safety: Volume II: Biological Agents and Methods for Industrial Waste Management*. Singapore: Bharagava, R.N., Saxena, pp. 53–76.

Alvarez, C.C., Bravo Gómez, M.E. and Hernández Zavala, A. (2021) Hexavalent chromium: Regulation and health effects. *Journal of Trace Elements in Medicine and Biology* 65, 126729. DOI: 10.1016/j.jtemb.2021.126729.

Ankit, B.K. and Korstad, J. (2022) Phycoremediation: Use of algae to sequester heavy metals. *Hydrobiology* 1(3), 288–303. DOI: 10.3390/hydrobiology1030021.

Aralbaeva, A.N., Yeszhanova, G.A., Aralbayev, A.N., Zhamanbayeva, G.T., Zhaparkulova, N.I. *et al.* (2024) Overview on the heavy metal toxicity mechanisms and the role of alimentary factors in detoxification. *International Journal of Biology and Chemistry* 17(2), 74–95. DOI: 10.26577/IJBCh2024v17.i2.7.

Bala, S., Garg, D., Thirumalesh, B.V., Sharma, M., Sridhar, K. *et al.* (2022) Recent strategies for bioremediation of emerging pollutants: A review for a green and sustainable environment. *Toxics* 10, 484. DOI: 10.3390/toxics10080484.

Balali-Mood, M., Naseri, K., Tahergorabi, Z., Khazdair, M.R. and Sadeghi, M. (2021) Toxic mechanisms of five heavy metals: Mercury, lead, chromium, cadmium, and arsenic. *Frontiers in Pharmacology* 12, 643972. DOI: 10.3389/fphar.2021.643972.

Barquilha, C.E.R., Cossich, E.S., Tavares, C.R.G. and Silva, E.A. (2017) Biosorption of nickel(II) and copper(II) ions in batch and fixed-bed columns by free and immobilized marine algae *Sargassum* sp. *Journal of Cleaner Production* 150, 58–64. DOI: 10.1016/j.jclepro.2017.02.199.

Charkiewicz, A.E., Omeljaniuk, W.J., Garley, M. and Nikliński, J. (2025) Mercury exposure and health effects: What do we really know? *International Journal of Molecular Sciences* 26(5), 2326. DOI: 10.3390/ijms26052326.

Chen, Z., Osman, A.I., Rooney, D.W., Oh, W.-D. and Yap, P.-S. (2023) Remediation of heavy metals in polluted water by immobilized algae: Current applications and future perspectives. *Sustainability* 15(6), 5128. DOI: 10.3390/su15065128.

Ciempiel, W., Czemierska, M., Wiącek, D., Szymańska, M., Jarosz-Wilkołazka, A. *et al.* (2025) Lead biosorption and chemical composition of extracellular polymeric substances isolated from mixotrophic microalgal cultures. *Scientific Reports* 15(1), 9093. DOI: 10.1038/s41598-025-94372-9.

Danouche, M., El Ghachtouli, N. and El Arroussi, H. (2021) Phycoremediation mechanisms of heavy metals using living green microalgae: Physicochemical and molecular approaches for enhancing selectivity and removal capacity. *Heliyon* 7(7), e07609. DOI: 10.1016/j.heliyon.2021.e07609.

Dey, S., Tripathy, B., Kumar, M.S. and Das, A.P. (2023) Ecotoxicological consequences of manganese mining pollutants and their biological remediation. *Environmental Chemistry and Ecotoxicology* 5, 55–61. DOI: 10.1016/j.enceco.2023.01.001.

Diaconu, L.I., Covaliu-Mierlă, C.I., Păunescu, O., Covaliu, L.D., Iovu, H. *et al.* (2023) Phytoremediation of wastewater containing lead and manganese ions using algae. *Biology* 12(6), 773. DOI: 10.3390/biology12060773.

Ding, M., Shi, S., Qie, S., Li, J. and Xi, X. (2023) Association between heavy metals exposure (cadmium, lead, arsenic, mercury) and child autistic disorder: A systematic review and meta-analysis. *Frontiers in Pediatrics* 11, 1169733. DOI: 10.3389/fped.2023.1169733.

Ganguly, A., Nag, S., Bhowmick, T.K. and Gayen, K. (2024) Bioremediation of chromium (VI) and arsenic (III) using isolated microalgal (*Chlorella thermophilia*): Analysis of growth, biomolecular compositions (carbohydrate, protein, chlorophyll) and biosorption kinetics. *Algal Research* 82, 103635. DOI: 10.1016/j.algal.2024.103635.

Ganie, S.Y., Javaid, D., Hajam, Y.A. and Reshi, M.S. (2023) Arsenic toxicity: Sources, pathophysiology and mechanism. *Toxicology Research* 13(1), tfad111. DOI: 10.1093/toxres/tfad111.

Garg, S.K., Tripathi, M. and Srinath, T. (2012) Strategies for chromium bioremediation of tannery effluent. *Reviews of Environmental Contamination and Toxicology* 217, 75–140. DOI: 10.1007/978-1-4614-2329-4_2.

Greeshma, K., Kim, H.-S. and Ramanan, R. (2022) The emerging potential of natural and synthetic algae-based microbiomes for heavy metal removal and recovery from wastewaters. *Environmental Research* 215(Pt 1), 114238. DOI: 10.1016/j.envres.2022.114238.

Hagvall, L., Pour, M.D., Feng, J., Karma, M., Hedberg, Y. *et al.* (2021) Skin permeation of nickel, cobalt and chromium salts in *ex vivo* human skin, visualized using mass spectrometry imaging. *Toxicology in Vitro* 76, 105232. DOI: 10.1016/j.tiv.2021.105232.

Haidar, Z., Fatema, K., Shoily, S.S. and Sajib, A.A. (2023) Disease-associated metabolic pathways affected by heavy metals and metalloid. *Toxicology Reports* 10, 554–570. DOI: 10.1016/j.toxrep.2023.04.010.

Hassan, S.H.A., Alomran, M.M., Alsugiran, N.I.A., Koutb, M., Ahmed, H. *et al.* (2025) Response surface optimization for cadmium biosorption onto the pre-treated biomass of red algae *Digenia simplex* as a sustainable indigenous biosorbent. *PeerJ* 13, e19776. DOI: 10.7717/peerj.19776.

Khan, M.T.A., Hassan, S.H.A., Al-Battashi, H. and Abed, R.M.M. (2025) Biosorption of hexavalent chromium using algal biomass: Isotherm and kinetic studies. *Algal Research* 85, 103867. DOI: 10.1016/j.algal.2024.103867.

Kharel, H.L., Jha, L., Tan, M. and Selvaratnam, T. (2024) Removal of cadmium (II) from aqueous solution using *Galdieria sulphuraria* CCMEE 5587.1. *Biotech (Basel, Switzerland)* 13(3), 28. DOI: 10.3390/biotech13030028.

Koh, H.Y., Kim, T.H., Sheen, Y.H., Lee, S.W., An, J. *et al.* (2019) Serum heavy metal levels are associated with asthma, allergic rhinitis, atopic dermatitis, allergic multimorbidity, and airflow obstruction. *Journal of Allergy and Clinical Immunology in Practice* 7(8), 2912–2915. DOI: 10.1016/j.jaip.2019.05.015.

Kumar, J.I.N. and Oommen, C. (2012) Removal of heavy metals by biosorption using freshwater alga *Spirogyra hyalina*. *Journal of Environmental Biology* 33(1), 27–31.

Kumar, A., Kumar, A., Cabral-Pinto, M.M.S., Chaturvedi, A.K., Shabnam, A.A. *et al.* (2020) Lead toxicity: Health hazards, influence on food chain, and sustainable remediation approaches. *International Journal of Environmental Research and Public Health* 17(7), 2179. DOI: 10.3390/ijerph17072179.

Kumar, R.M., Franklin, J. and Raj, S.P. (2013) Accumulation of heavy metals (Cu, Cr, Pb and Cd) in freshwater micro algae (*Chlorella* sp.). *Journal of Environmental Science and Engineering* 55(3), 371–376.

Lin, H.C., Liu, Y.J. and Yao, D.J. (2024) Preparation of magnetic microalgae composites for heavy metal ions removal from water. *Heliyon* 10(18), e37445. DOI: 10.1016/j.heliyon.2024.e37445.

Liu, C., Wen, X., Pan, H., Luo, Y., Zhou, J. *et al.* (2024) Bioremoval of Co (II) by a novel halotolerant microalgae *Dunaliella* sp. FACHB-558 from saltwater. *Frontiers in Microbiology* 15, 1256814. DOI: 10.3389/fmicb.2024.1256814.

Luka, Y., Highina, B.K., Zubairu, A., Adeleke, A.J., Hamadou, M. *et al.* (2024) Biosorption as technique for remediation of heavy metals from wastewater using microbial biosorbent. *Biological Sciences* 04(1), 564–574. DOI: 10.55006/biolsciences.2024.4105.

Masindi, V. and Muedi, K.L. (2018) Environmental contamination by heavy metals. *Heavy Metals* 10(4), 115–133. DOI: 10.5772/intechopen.76082.

Oyebamiji, O.O., Corcoran, A.A., Navarro Pérez, E., Ilori, M.O., Amund, O.O. *et al.* (2021) Lead tolerance and bioremoval by four strains of green algae from Nigerian fish ponds. *Algal Research* 58, 102403. DOI: 10.1016/j.algal.2021.102403.

Parmar, K.S. and Patel, K.M. (2025) Biosorption and bioremediation of heavy metal ions from wastewater using algae: A comprehensive review. *World Journal of Microbiology and Biotechnology* 41(7), 262. DOI: 10.1007/s11274-025-04424-5.

Saini, S. and Dhania, G. (2020) Cadmium as an environmental pollutant: Ecotoxicological effects, health hazards, and bioremediation approaches for its detoxification from contaminated sites. In: *Bioremediation of Industrial Waste for Environmental Safety: Volume II: Biological Agents and Methods for Industrial Waste Management.* pp. 357–387. DOI: 10.1007/978-981-13-3426-9_15.

Salama, E.S., Roh, H.S., Dev, S., Khan, M.A., Abou-Shanab, R.A.I. *et al.* (2019) Algae as a green technology for heavy metals removal from various wastewater. *World Journal of Microbiology and Biotechnology* 35(5), 75. DOI: 10.1007/s11274-019-2648-3.

Salimi, A., Ghanbarizadeh, P., Mirvakili, A. and Moheimani, N.R. (2025) Optimizing heavy metal remediation of synthetic wastewater using *Chlorella vulgaris* and *Sargassum angustifolium*: A comparative analysis of biosorption and bioaccumulation techniques. *Science of the Total Environment* 992, 179938. DOI: 10.1016/j.scitotenv.2025.179938.

Sbihi, K., Elhamji, S., Lghoul, S., Aziz, K., El Maallem, A. *et al.* (2024) Biosorption of hexavalent chromium by freshwater microalgae *Craticula subminuscula* from aqueous solutions. *Sustainability* 16(2), 918. DOI: 10.3390/su16020918.

Shamim, S. (2018) Biosorption of heavy metals. *Biosorption* 2, 21–49. DOI: 10.5772/intechopen.72099.

Shamshad, I., Khan, S., Waqas, M., Ahmad, N., Ur-Rehman, K. *et al.* (2015) Removal and bioaccumulation of heavy metals from aqueous solutions using freshwater algae. *Water Science and Technology* 71(1), 38–44. DOI: 10.2166/wst.2014.458.

Singh, V.P., Godara, P. and Srivastava, A. (2024) Sustainable microalgal bioremediation of heavy metals and dyes from synthetic wastewater: Progressing towards united nations sustainable development goals. *Waste Management Bulletin* 2(4), 123–135. DOI: 10.1016/j.wmb.2024.10.005.

Spain, O., Plöhn, M. and Funk, C. (2021) The cell wall of green microalgae and its role in heavy metal removal. *Physiologia Plantarum* 173(2), 526–535. DOI: 10.1111/ppl.13405.

Tenza, N.P., Schmidt, S. and Mahlambi, P.N. (2025) Unlocking the potential of *Chlorella* sp. biomass: An effective adsorbent for heavy metals removal from wastewater. *Frontiers in Environmental Chemistry* 6, 1–15. DOI: 10.3389/fenvc.2025.1531726.

Tizabi, Y., Bennani, S., El Kouhen, N., Getachew, B. and Aschner, M. (2023) Interaction of heavy metal lead with gut microbiota: Implications for autism spectrum disorder. *Biomolecules* 13(10), 1549. DOI: 10.3390/biom13101549.

Torres, E. (2020) Biosorption: A review of the latest advances. *Processes* 8(12), 1584.

Tripathi, M., Vikram, S., Jain, R.K. and Garg, S.K. (2011) Isolation and growth characteristics of chromium(VI) and pentachlorophenol tolerant bacterial isolate from treated tannery effluent for its possible use in simultaneous bioremediation. *Indian Journal of Microbiology* 51, 61–69. DOI: 10.1007/s12088-011-0089-2.

Tripathi, M., Pathak, S., Singh, R., Singh, P., Singh, P.K. *et al.* (2024). Adsorptive remediation of hexavalent chromium using agro-waste rice husk: Optimization of process parameters and functional groups characterization using FTIR analysis. *The Scientific Temper* 15(04), 2971–2976. DOI: 10.58414/scientifictemper. 2024.15.4.02.

Tripathi, M., Singh, R., Lal, B., Haque, S., Ahmad, I. *et al.* (2025) Marine microbial bioremediation of heavy metal contaminants in waste water for health and environmental sustainability: A review. *Indian Journal of Microbiology* 65(2), 573–582. DOI: 10.1007/s12088-024-01427-y.

Utomo, H.D., Tan, K.X.D., Choong, Z.Y.D., Yu, J.J., Ong, J.J. *et al.* (2016) Biosorption of heavy metal by algae biomass in surface water. *Journal of Environmental Protection* 07(11), 1547–1560. DOI: 10.4236/jep.2016.711128.

Vellingiri, B., Suriyanarayanan, A., Selvaraj, P., Abraham, K.S., Pasha, M.Y. *et al.* (2022) Role of heavy metals (Cu), (As), (Cd), (Fe) and (Li)) induced neurotoxicity. *Chemosphere* 301, 134625. DOI: 10.1016/j.chemosphere.2022.134625.

Ye, J., Xiao, H., Xiao, B., Xu, W., Gao, L. *et al.* (2015) Bioremediation of heavy metal contaminated aqueous solution by using red algae *Porphyra leucosticta. Water Science and Technology: A Journal of the International Association on Water Pollution Research* 72(9), 1662–1666. DOI: 10.2166/wst.2015.386.

Zeng, G., He, Y., Liang, D., Wang, F., Luo, Y. *et al.* (2022) Adsorption of heavy metal ions copper, cadmium and nickel by *Microcystis aeruginosa. International Journal of Environmental Research and Public Health* 19(21), 13867. DOI: 10.3390/ ijerph192113867.

Zhao, C., Liu, G., Tan, Q., Gao, M., Chen, G. *et al.* (2023) Polysaccharide-based biopolymer hydrogels for heavy metal detection and adsorption. *Journal of Advanced Research* 44, 53–70. DOI: 10.1016/j.jare.2022.04.005.

Znad, H., Awual, M.R. and Martini, S. (2022) The utilization of algae and seaweed biomass for bioremediation of heavy metal-contaminated wastewater. *Molecules (Basel, Switzerland)* 27(4), 1275. DOI: 10.3390/molecules27041275.

Recent Approaches for Enhancing Microbial Performance for Remediation of Heavy Metals from Wastewater

5

5.1 Introduction

Heavy metal (HM) pollution is a major global concern and poses serious risks to aquatic ecosystems, terrestrial ecosystems and human health through water and soil contamination and food chain entry. Urbanization and industrial developments have increased HM concentrations in the environment beyond their permissible limits. Various sectors such as mining, metallurgical industries, electroplating, agricultural runoff and urban wastewater are contributing to HM presence in the environment. HMs such as cadmium (Cd), lead (Pb), chromium (Cr), mercury (Hg) and arsenic (As) are non-degradable and bioaccumulate through food webs, causing ecological damage and serious human health impacts (Sarker *et al.*, 2023). HMs enter aquatic ecosystems and agricultural soils mainly through industrial effluents, irrigation with polluted water and the overapplication of fertilizers and pesticides (Gupta *et al.*, 2021).

Conventional physical and chemical methods such as precipitation, ion exchange and adsorption on activated carbonate can be effective, but are costly, labour-intensive and risk secondary pollution. These problems have intensified the need for eco-friendly solutions such as biological approaches. HMs and toxic substances can be eliminated from the environment by using biological approaches such as bioremediation, phytoremediation, bioaugmentation and engineered consortia. Bioremediation, using hyperaccumulators, beneficial microbes and genetically modified organisms, offers a sustainable and eco-friendly alternative but requires a deep mechanistic understanding to be an effective strategy (Zha *et al.*, 2024). Recently, advancements in the fields of molecular biology and synthetic and nanotechnology have dramatically improved the performance and utility of microbial bioremediation in the decantation of HMs (Thai *et al.*, 2023).

Corresponding author: sukhminderjit.uibt@cumail.in

Various interlinked frontiers have emerged over the last few years that promise step changes in microbial remediation performance. Genetic engineering and high-throughput omics approaches such as genomics, transcriptomics, proteomics and metabolomics are being studied to understand the mechanisms involved in improving metal tolerance, sequestration and enzymatic transformation in bacterial, fungal and algal strains, enabling targeted uptake and even biosensor-guided detoxification. Synthetic biology has opened new horizons in microbial bioremediation by providing innovative tools to engineer microorganisms with superior traits. These include rapid detection using HM biosensors and efficient detoxification of HMs, alongside improved tolerance to toxic environments. Synthetic biology adopts engineering principles to design and construct synthetically engineered microorganisms, thereby enabling enhanced tolerance against toxic chemicals for better bioremediation (Thai *et al.*, 2023; Tiwary, 2025). New designs of robust consortia and nanomaterials are being explored to improve sorption/selectivity, to deliver functional groups and to create synergistic remediation platforms (De Silva *et al.*, 2025; Qattan, 2025).

This chapter provides a comprehensive overview of various advanced approaches such as the application of genetically engineered microorganisms, various omics techniques, nanotechnology and synthetic biology in the bioremediation of HMs.

5.2   Emerging Genetic Approaches for Microbial Enhancement

5.2.1   Genetically engineered microbes

In recent years, advances in genetic engineering have significantly expanded the potential of various microbes for HM bioremediation. Microbes (bacteria, fungi, yeast and algae) use their genetic and metabolic adaptations to detoxify HMs. The natural efficiency of microbes to tolerate and accumulate toxic metals is usually limited and regulated by various mechanisms. By introducing or overexpressing genes in for example metallothioneins (MTs), phytochelatins and mer operons, scientists have created genetically engineered microorganisms with enhanced adsorption, transformation and detoxification abilities by enhancing gene expression linked to HM resistance. Various genetic engineering strategies at the gene and genome levels have been utilized to insert, delete or replace nucleotides within bacterial DNA fragments to create genetically engineered strains. However, their effectiveness depends on both microbial genetics and environmental conditions, while large-scale application faces practical challenges (Naiel *et al.*, 2024). Using genetically modified microbes for bioremediation requires a thorough understanding of the metabolic pathways of the microorganisms.

A recombinant *Escherichia coli* DH5α strain was constructed to display MT yeast CUP1 on its surface using the Lpp-OmpA system. MTs are small and cysteine-rich, metal-binding proteins that take part in metal metabolism and antioxidation. The constructed *E. coli* strain has not only improved Cd^{2+} resistance but also achieved a removal efficiency of 95.2% with sulfhydryl and sulfonyl groups contributing to the Cd binding. Overall, surface-displayed MTs can have great potential for HM remediation (He *et al.*, 2025).

In another example, the MT gene *EGR_09832* from *Echinococcus granulosus* was fused with Lpp-OmpA to place MT on the surface of *E. coli* DH5α. The engineered strain showed an increased Cd tolerance to levels of 160 mg/l and also had more than two times the adsorption capacity in comparison with the wild type. These data show that the use of surface-displayed Cd-binding peptides may significantly increase microbial Cd bioremediation capacity (He *et al.*, 2024).

Wang *et al.* (2024a) showed the use of genetically engineered *Deinococcus radiodurans*, an extremophile, as a beneficial microbial inoculant for reducing HM accumulation and toxicity in crops. *D. radiodurans* was genetically transformed to express an Lpp-OmpA protein with two metal-binding domains (PbBD and MTT5) on the outer membrane. The genetically modified strain, LOPM, exhibited a radical increase in metal tolerance and accumulated 4.9-fold more Cd and 3.2-fold more Pb than the wild type. When colonizing rice roots, LOPM reduced Cd accumulation by 47.0% in roots and 43.4% in shoots, and Pb accumulation by 55.4% in roots and 26.9% in shoots. In addition to demonstrating an impact on HM uptake, LOPM reduced oxidative stress from Cd and Pb in rice by lowering levels of reactive oxygen species (ROS) and increasing antioxidant enzymatic activity.

Xue *et al.* (2024) integrated an *arsR* gene into *Pseudomonas putida* KT2440 including a programmable suicide circuit to regulate growth and maintain biosafety in bioremediation of As. The engineered *P. putida* KT2440 strain, along with biochar, facilitated the removal of As from mining wastewater, and detoxification was proven to be effective up to 20 mg/l.

A strain with high mercury-removal capacity and effective tolerance was created by the introduction of the mer operon into *Shewanella oneidensis* bacteria. The engineered strain exhibited broad resistance to Hg and good removal ability. Particles were seen on the bacterial surface from remediation, and further analysis showed the engineered bacteria were able to reduce Hg^{2+} to Hg^0 (Fang *et al.*, 2024). *S. azerbaijanica* was engineered for enhanced metal bioremediation by cloning the *mtrC* gene. The recombinant bacterium's bioremediation efficiency requires further investigation (Rastkhah *et al.*, 2024).

Saccharomyces cerevisiae was engineered by introducing four Cd^{2+} peptides genes which were identified using bacteriophage display peptide library and cell surface display technologies, via the pYD1 plasmid. The engineered yeasts showed significantly higher Cd^{2+} adsorption than the control strain, achieving 35% greater efficiency. Furthermore, immobilization in sodium alginate enhanced adsorption by 55.7% compared with the control (Wang *et al.*, 2024c).

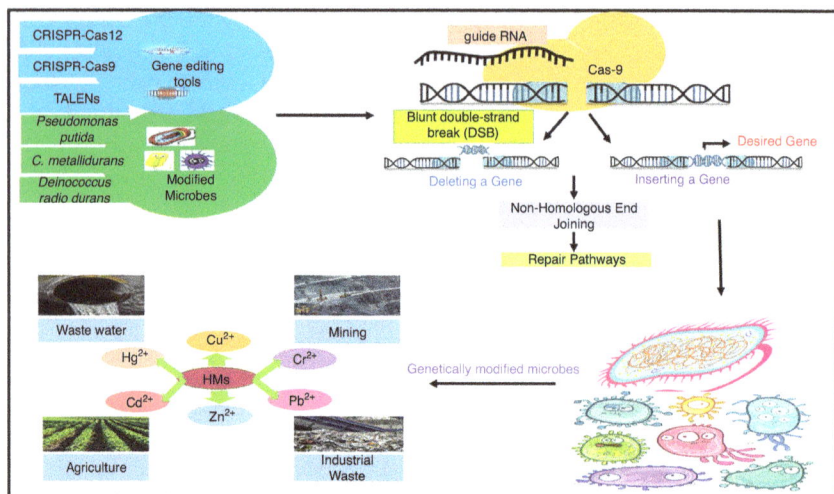

Fig. 5.1. Gene-edited microbes for bioremediation of heavy metals (HMs) from diverse sources.

Recombinant *E. coli* expressing *Tetrahymena* MTT5 fused with Lpp-OmpA in its outer membrane effectively removed Cd from mixed-metal wastewater. The recombinant strain showed 4.9-fold higher Cd adsorption than wild-type strains, highlighting its potential for field applications (Lu *et al.*, 2023).

A recombinant native-form *Sinopotamon henanense* metallothioneins (ShMT), present in the freshwater crab *S. henanense*, was successfully expressed in *E. coli* and purified using immunoaffinity chromatography. Metal-binding assays revealed strong affinity for copper (Cu), Cd and zinc (Zn), with binding preference ranked as Cu > Cd > Zn. This strategy enables production of functional ShMT and provides new insights into its metal-binding properties and potential biological roles (He *et al.*, 2019).

5.2.2   Genome-editing tools

Gene editing has become more popular because it can provide precise modifications to genomes within microbes of interest, without foreign genes being introduced. The most common tools to engineer beneficial microbes are TALENs (transcription activator-like effector nucleases), zinc finger nucleases, CRISPR (clustered regularly interspaced short palindromic repeats) Cas9 and Cas12 (Fig. 5.1). These technologies are primarily applied to regulate the expression of target genes, either by enhancing or suppressing their activity. Among these, CRISPR-Cas - has recently emerged as a highly efficient and user-friendly gene-editing tool (Liu *et al.*, 2021; Kavya *et al.*, 2024). Over the past decade, CRISPR-Cas technologies have transformed biological research whereby targeted microbial genomes can be precisely and programmably edited to enhance metal-uptake, metal-tolerance and metal-detoxification mechanisms. Using CRISPR-Cas systems, researchers have knocked out genes for metal efflux to enhance intracellular retention of toxic ions as well as introducing genes encoding high-affinity metal-binding proteins including MTs, polyhistidine-rich peptides and phytochelatin synthases. The ability to rapidly modify genomes is accelerating the mapping of genotype–phenotype relationships and the design of new, desired biological traits. CRISPR/CRISPR-derived tools have been used to edit microbes, with implications for HM removal by resistance, biosorption, biomineralization and electron transfer (Burbano *et al.*, 2024). Advances in editing approaches like base editing and prime editing enable further adjustment of transporter specificity, enzyme activity and regulatory elements on a small scale without causing a double-strand break and reducing off-target effects. CRISPR-edited strains of *Shewanella oneidensis* exhibit a 1.8-fold greater reduction of Cr(VI) to Cr(III) relative to wild-type strains owing to amplified expression of chromate reductases and splayed electron transport chains (Chourey *et al.*, 2006; Jiang *et al.*, 2020).

Turco and co-workers (2022) developed a CRISPR-Cas9 system for genome editing in *Cupriavidus metallidurans* CH34, with optimization of sgRNA and Cas9 expression to reduce toxicity and enhance efficiency. A comprehensive library of constitutive promoters and a hybrid inducible P_{BAD}–riboswitch system was characterized for precise gene regulation. This represents the first extensive set of regulatory elements for CH34, and the CRISPR-Cas9 tool demonstrated robust performance, yielding recombinant clone isolation efficiencies of between 2% and 100%.

Liang *et al.* (2020) established a CRISPR-Cas9 system that increased transformation using bacteriophages in *Rhodococcus* sp. 89-fold, and increased editing efficiency to 75% using Che9c60/61 recombinases. A triple-plasmid recombineering approach enabled efficient gene deletion, insertion and mutation. Using this system, an engineered strain THY enhanced acrylamide production from 405 to 500 g/L and reduced by-products.

Chen *et al.* (2022) developed and validated Cas12 systems for efficient genome editing in *S. oneidensis* MR-1, an electroactive bacterium widely used for Cr(VI) and other metal reductions. Targeted manipulation of genes linked to extracellular electron transfer was done with low toxicity and high editing efficiency in AT-rich regions. This study established CRISPR-AsCpf1 and CRISPR-BhCas12b systems well suited for AT-rich regions. BhCas12b enabled gene deletion (41.7%), promoter replacement (25%) and high-efficiency gene insertion (94.4%), while AsCpf1 achieved 83.9% insertion efficiency.

In parallel, genome-wide DNA-binding technologies such as TALENs, which recognize host-specific nucleotide sequences, have also been applied (Sun and Zhao, 2013).

5.3   Omics Technologies in Heavy Metal Remediation

Omics technologies like genomics, transcriptomics, proteomics and metabolomics are emerging techniques used for environment management specifically
in the detoxification of HMs through bioremediation (Table 5.1). These techniques provide in-depth molecular detail about microbes that live in particular
niches with mechanisms that can be applied in bioremediation. These developing technologies show a reduction in time spent exploring specific niches and
have become a replacement for the antiquated culture-based methods used to
establish more useful microbiomes (Lata *et al.*, 2023). The following sections
include an in-depth discussion of various omics methods for the bioremediation of HMs. Applications of integrated multi-omics methods further deepen
mechanistic understanding and have provided advanced insights into the
remediation and detoxification of toxic HMs. HMs disrupt microbial cellular
homeostasis by binding to protein active sites, which alters enzyme activity
and metabolic pathways, ultimately impairing their growth and survival.
Genomics and transcriptomics identify candidate genes and their expression
patterns in normal and stress conditions. Proteomics and metabolomics
measure molecules such as enzymes, transporters, metal-binding proteins and
stress–response complexes that perform detoxification, emphasizing the need
for linking genotype to phenotype in metal-stressed systems (Fig. 5.2).

5.3.1   Metagenomics and genomics: identifying metal resistance genes

Metagenomics has emerged as a powerful tool for exploring novel microbial
communities in addressing environmental contamination, as it operates
largely independent of traditional culturing techniques. Through this
approach, suitable and efficient microorganisms can be identified by studying
their whole genome for remediation purposes. It enables rapid analysis of
genome sequences of single microorganisms or of whole microbiomes associated with an environmental niche, associated metabolites and the detoxification potential of microbes. Since many environmentally relevant microbes
are difficult to culture under laboratory conditions (Rashid and Stingl, 2015),
metagenomics provides access to their DNA sequences, offering valuable
insights into their roles in bioremediation and supporting the development of
advanced remediation strategies.

Several genetic determinants responsible for metal resistance have
been characterized in diverse bacterial phyla (Gillieatt and Coleman, 2024).
HM resistomes are highly diverse, and metal resistance genes (MRGs) are
widespread among the microbial community, with about 86% of complete
genomes carrying potential MRGs (Pal *et al.*, 2014). MRGs may play roles in
maintaining the homeostasis of essential metals, and providing resistance
at higher metal concentrations. These datasets enable resistome mapping,
the discovery of novel metal-binding proteins and the design of synthetic

Table 5.1. Multi-omics approaches in heavy metal bioremediation.

Heavy metal	Omic approach	Findings	Reference
Cadmium (Cd)	Whole-genome sequencing	Revealed a 7.32 Mb genome with 66.39% GC content and 6504 protein-coding genes, including those linked to nutrient cycling, hormone regulation, stress tolerance and pathogen antagonism. Functional assays using AAS, FTIR and SEM-EDS confirmed its ability to adsorb Cd^{2+} via surface groups such as C–O–C, $P=O$ and O–H	Feng *et al.*, 2025
Cr(VI)	Comparative metagenomics + metatranscriptomics	Identified six novel Cr-tolerance/remediation genes from uncultured microbes; two genes (*mcr*, *gsr*) expressed in *E. coli* had a drop of c.50% Cr(VI) from industrial wastewater (200–600 μM) for 17 days	Pei *et al.*, 2020
Cr(VI)	Metagenomics + metatranscriptomics (shotgun RNA-seq)	Metagenomes were similar between control and stress conditions Metatranscriptomes revealed that genes associated with oxidative stress, Cr(VI) transport, resistance and reduction, as well as genes with unknown functions, were 2–10 times upregulated after Cr(VI) treatment. in *E. coli* MT results also revealed an increase in the expression of some rare genera (at least two times) after Cr(VI) treatment	Yu *et al.*, 2021
Silver (As)	Differential proteomics (2D/MS)	Changes in proteins linked to antioxidative defense, stress chaperones and enzymes involved in sulfur metabolism, indicating microbial-assisted mitigation via enhanced ROS scavenging and S-metabolism	Alka *et al.*, 2021
Cadmium (Cd)	Quantitative proteomics (label-free LC-MS/MS)	Upregulation of various proteins such as metal-binding proteins, efflux-related proteins and stress enzymes in *Enterobacter* sp. FM-1	Li *et al.*, 2023

Continued

Table 5.1. Continued

Heavy metal	Omic approach	Findings	Reference
Lead (Pb)	Global proteomics (LC-MS/MS)	Identified 43 proteins that showed statistically significant changes in abundance under Pb exposure (24 upregulated and 19 downregulated: ANOVA, $P \leq 0.05$; fold change ≥ 1.5), categorized into different functional categories – metabolism, cellular processes and signalling, and information storage and processing – in *Penicillium chrysogenum*	Algahmadi et al., 2024
Silver (As)	Quantitative proteomics	Quantitative proteomics showed overproduction of the SoxAXYZB complex, along with electron carriers cytochrome c551/c5 and HiPIP III. EPR analysis indicated their role in reducing the photosynthetic reaction centre, supporting a model where biologically produced thioarsenates are oxidized by the Sox system, with electrons funnelled to the reaction centre in photosynthetic bacterium	D'Ermo et al., 2024
Lead (Pb), cadmium (Cd)	Metagenomics and metabolomics	Main functional microbes *Janibacter* sp., *Lysobacter* sp., *Ornithinimicrobium* sp., *Bacillus* sp. and *Salinimicrobium* sp. were identified. Synergistic remediation significantly increased available soil phosphate content and decreased extractable Pb^{2+} and Cd^{2+}. Four HM resistance genes (*ZitB*, *czcD*, *zntA* and *cmtR*) were identified	Zhang et al., 2025b
Chromium (Cr)	Transcriptomics and metabolomics	*Bacillus safensis* BSF-4 halophilic bacterium reduced Cr(VI) efficiently (89%) in 72 hour. Upregulation of various pathways such as membrane-associated transport systems and quorum sensing for stress adaptation was observed	Liu et al., 2025

2D/MS, two-dimensional mass spectrometry; AAS, atomic absorption spectroscopy; ANOVA, analysis of variance; EPR, electron paramagnetic resonance; FTIR, Fourier transformer infrared spectroscopy; HM, heavy metal; LC-MS/MS, liquid chromatography–mass spectrometry/mass spectrometry; MT, metallothionein; ROS, reactive oxygen species; SEM-EDS, scanning electron microscopy–energy dispersive spectroscopy.

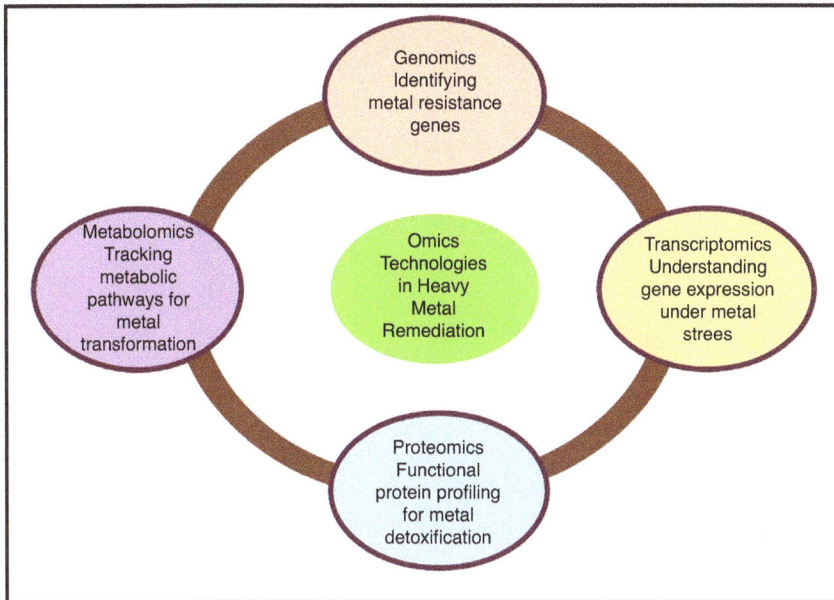

Fig. 5.2. Role of omics approaches in heavy metal remediation.

or enriched consortia tailored for specific contaminants and geochemical contexts. Metagenomic case studies from mining-impacted soils and wastewater show enrichment of metal-resistance operons and mobilome elements that help predict community resilience (Galisteo *et al.*, 2024).

Recent studies have reported that desertification, mining and industrial activities have shaped their microbial communities. Shotgun metagenomic sequencing showed a balanced presence of archaea and bacteria, dominated by Methanobacteriota, Pseudomonadota, Bacteroidota, Gemmatimonadota and Balneolota, in environmental samples showing adaptations for HM resistance via efflux systems such as ZntA and CopA, and As detoxification primarily by Methanobacteriota (Galisteo *et al.*, 2024). By carrying out various genomic studies, different MRGs have been identified and some of the genes responsible for specific metal removal are discussed here.

To detoxify cytoplasmic Cu, Gram-negative bacteria typically encode at least one Cu^+-ATPase gene, and strains showing pathogenic and symbiotic association usually carry extra copies. In *Pseudomonas aeruginosa*, two homologous ATPases, CopA1 and CopA2, both export Cu to the periplasm but serve distinct roles: CopA1 regulates cytoplasmic Cu^+ levels, while CopA2 likely facilitates Cu delivery to cytochrome c oxidase. Overall, ATPase-driven Cu efflux is the primary detoxification strategy, complemented by periplasmic systems such as multicopper oxidases, metallochaperones and resistance–nodulation–division (RND) transporters (Bondarczuk and

Piotrowska-Seget, 2013). *E. coli* maintains Cu homeostasis through CopA, CueO and the CusCFBA system, while some strains harbour plasmid-encoded *pco* genes for additional periplasmic detoxification. Cu regulation involves the MerR-like and two-component systems (Rensing and Grass, 2003). Similarly, the pco system, specific to γ-Proteobacteria, provides resistance to Cu (Staehlin *et al.*, 2016), while the tcr determinant, found in *Enterococcus* sp., also contributes to Cu tolerance (Hasman and Aarestrup, 2002).

The mer operon encodes Hg resistance and is widely distributed among Gram-positive and Gram-negative bacteria. The core HgR operon typically encodes MerA, a mercuric reductase that reduces toxic Hg(II) to inert Hg(0), and MerT, a membrane protein mediating Hg(II) uptake, both regulated by the metal-responsive factor MerR. Various operons will also include MerB, which degrades organomercurials, and additional proteins related to transport, among other functions. These genes are found on chromosomes, plasmids and transposons, sometimes with different and repeated permutations. This genetic mobility protects bacteria from mercury toxicity, and also ultimately affects mercury cycling, and has bioremediation possibilities (Barkay *et al.*, 2003).

The ars operon is responsible for arsenic and antimony resistance, largely through chemical modification and efflux, within Actinomycetota, Bacillota, Chloroflexi (CFB) group and Pseudomonadota (Ben Fekih *et al.*, 2018).

The chr operon confers efflux-mediated resistance to chromium in the Pseudomonadota, Actinomycetota, Bacillota and CFB group (Branco *et al.*, 2008). The cnr determinant also provides efflux-mediated resistance to cobalt (Co) and nickel (Ni) in Pseudomonadota and Planctomycetes (Marrero *et al.*, 2007). The cus system confers efflux-mediated resistance to silver (Ag) and Cu in Pseudomonadota (Staehlin *et al.*, 2016). The czc system confers resistance to Cd, Co and Zn in the CFB group, and in Pseudomonadota (Nies, 1995). Conversely, the czr system confers resistance to Cd and Zn in Actinobacteria, Bacillota and Pseudomonadota (Hassan *et al.*, 1999). Similarly, the ncc system provides Ni, Co and Cd resistance in Pseudomonadota (Schmidt and Schlegel, 1994).

In *Pseudomonas* it is expected that the sil operon, as described by Gupta *et al.* (1999), should confer resistance to Ag, with the *znt* gene presumably responsible for Zn homeostasis. The pbr operon provides Pb resistance with efflux and sequestration in Bacillota and *Pseudomonas* sp. (Borremans *et al.*, 2001).

5.3.2 Metatranscriptomics and transcriptomics: exploring gene expression during heavy metal stress

Metatranscriptomics and transcriptomics are the molecular approaches that use RNA sequencing of microorganisms to reveal molecular responses to different environments. These approaches analyse the gene expression through RNAs under stress conditions in microorganisms and characterize responses to stress, thus helping to characterize the active microorganisms associated with their active genes in specific environmental conditions (Simon and Daniel, 2011; White *et al.*, 2017). Transcriptomic approaches can provide genomic-scale resolution of differentially expressed genes (DEGs) under various levels of metal stress that can allow the characterization of important pathways associated with adaptation such as transport, detoxification, ROS and transcription factors.

Functional protein profiling for metal detoxification using metagenomic mRNA from environmental samples allows the study of the responses of microbial communities to HM stress (Yu *et al.*, 2021). Functional metatranscriptomics helps the identification of genes that allow microorganisms to cope with extreme conditions in the environment. In metatranscriptomics, all the RNA from environmental samples is extracted and sequenced with various intermediate steps such as construction of a cDNA library and screening for HM-tolerant transcripts (Thakur *et al.*, 2018; Mukherjee and Reddy, 2020). Thakur *et al.* (2018) employed this approach and screened a cDNA library from a contaminated Cd site and discovered a yeast transformant that exhibited significant tolerance to multiple stresses including high concentrations of Co, Cu and Zn.

Lehembre *et al.* (2013) utilized functional metatranscriptomics to study the roles of soil microbiota in HM resistance. Their work utilized a soil metatranscriptome library; which comprised total RNA from a diverse range of soil microbes, and screened this to determine if the library imparted increased HM resistance to Cd- or Zn-sensitive haploid yeast mutants. They then identified several proteins previously not associated with HM resistance, but involved in metal detoxification, including: (i) BolA proteins, which are involved in regulating the transition from exponential growth to stationary phase and/or biofilm formation; (ii) saccharopine dehydrogenase (involved in Zn tolerance); and (iii) the C-terminal region of aldehyde dehydrogenase (ADH), which was implicated in Cd tolerance. In another study, Mukherjee *et al.* (2019) also utilized a eukaryotic cDNA library grown in HM-contaminated soil and recovered a clone of interest (PLCc38) which was homologous to ADH and exhibited tolerance to multiple metals (Cu, Cd, Zn and Co). ADH enzymes are key for degrading harmful aldehydes produced by abiotic stresses such as exposure to HMs.

Yu *et al.* (2021) investigated the short-term response of soil microbiota to Cr^{6+} stress (1 mM for 30 min) using a combined metagenomic and metatranscriptomic approach. While metagenomic analysis showed that 99% of microbial genera were shared between the control and stressed samples, metatranscriptomics revealed that 83% of microbes altered their RNA expression in response to Cr^{6+} exposure. Upregulated genes included those related to oxidative stress management, metal transport, resistance and Cr^{6+} reduction. Functional validation in *E. coli* confirmed that two previously uncharacterized upregulated genes contributed to Cr^{6+} remediation. In another study, Pei *et al.* (2020) employed comparative metagenomics and metatranscriptomics on long-term Cr^{6+}-contaminated riparian soils, identifying six novel Cr^{6+}-tolerance genes.

Expression of two of these genes, *mcr* and *gsr*, in *E. coli* resulted in nearly 50% reduction of Cr^{6+} from industrial wastewater (200–600 µM) within 17 days.

5.3.3 Proteomics: functional protein identification for heavy metal elimination

Proteomics, a core branch of omics, provides valuable information on dynamic protein functions that are vital to understanding mechanisms in cells. It involves the identification, characterization and functional classification of proteins within biological systems. The field also examines post-translational modifications, protein quantification and structural alterations triggered by internal or external stimuli. The process begins with peptide purification through extraction of total proteins, followed by separation via gel electrophoresis using fluorescent dyes or radioactive probes. The distinct proteins are then identified through mass spectrometry (Kaur *et al.*, 2023).

The application of proteomics in the bioremediation of HMs will provide insights into the functions, structures, localization and interactions of proteins, as well as their roles under normal and stress conditions. Through such studies, potential protein markers can be identified, where variations in their abundance correlate with physiological traits that indicate a genotype's level of stress tolerance. Proteomic profiling validates translation of transcripts and directly detects enzymes and binding proteins that mediate detoxification, complexation or biomineralization. Proteome profiling of metal-resistant microbes has reported the importance of MTs and other metal-binding proteins involved in detoxification. MTs occur across diverse organisms, exhibiting isoform-specific regulation and forming distinct metal-dependent 'metalloform' landscapes. Advanced chemoproteomic and native-MS approaches have mapped MT metal-occupancy states, providing insights into protein stability and function (Peris-Díaz *et al.*, 2024).

Various detoxification strategies have been clarified by proteomic studies such as the role of MT and metal-binding proteins, efflux and sequestration systems, antioxidant enzymes and metallochaperones and mapped dynamic regulation via phosphorylation. Integrated with other omics, proteomics has revealed the mechanistic networks underlying tolerance and remediation potential. Proteomic surveys consistently highlight the upregulation and post-translational activation of antioxidant and redox enzymes such as superoxide dismutase, catalase, peroxiredoxins, and glutathione S-transferases, emphasizing ROS detoxification as a key protective mechanism in plants, fungi and bacteria (Li *et al.*, 2023). Capdevila *et al.* (2024) reported that transporters and efflux systems, including P-type ATPases, cation diffusion facilitator (CDF) family proteins, ATP-binding cassette (ABC) transporters and metallochaperones, are enriched in microbes thriving in HM-contaminated sites underpinning strategies for efflux and vacuolar sequestration in both eukaryotic and prokaryotic systems. Furthermore, phosphoproteomics and post-translational modification-focused studies demonstrate that signalling cascades and phosphorylation events rapidly reprogramme the proteome under metal stress, modulating transporter trafficking, enzymatic activity and transcription factor regulation (Tang *et al.*, 2024).

5.3.4 Metabolomics: tracking metabolic pathways for heavy metal transformation

Metabolomics has emerged as a powerful tool for deciphering cellular processes, supported by standardized extraction methods and advanced analytical techniques for capturing the small-molecule currencies. Since metabolites closely reflect phenotypes by integrating gene expression, protein interactions and regulatory pathways, this approach provides deeper biological insights. Recent advances, including computational modelling, flux balance analysis (FBA), OptKnock and network-based tools, now enable precise prediction of microbial metabolic capabilities for biotechnological applications (Rani and Sagar, 2024; Thakur *et al.*, 2025). Microbial and plant communities transform HMs via reduction/oxidation, chelation, methylation/demethylation, biomineralization and volatilization. In HM remediation, metabolomics link genes and enzymes to these transformations *in situ*, hence revealing the role of different molecules such as organic acids, siderophores and methyl donors involved in mobility of metals. Recent reviews across soils, sediments and engineered systems have emphasized the importance of metabolomics alongside metagenomics and transcriptomics as a core pillar to investigate HM bioremediation (Tang *et al.*, 2024; Wang *et al.*, 2024d; Mohammad *et al.*, 2025). Metabolomics leverages advanced technologies – including solid-state NMR (nuclear magnetic resonance), LC-MS (liquid chromatography–mass spectrometry), CE-MS (capillary electrophoresis–mass spectrometry), GC-MS (gas chromatography–mass spectrometry) and other mass spectrometry methods – to analyse small molecules at a molecular level. Each technique offers different capabilities for detecting metabolite types and quantities, requiring tailored approaches to effectively study specific metabolic profiles (Eshawu and Ghalsasi, 2024).

Through metabolic profiling, researchers can pinpoint specific biomarkers and disrupted pathways linked to metal-induced toxicity. Such insights enable the design of targeted strategies to reduce harmful effects. Integrating metabolomics with other omics tools enhances the precision of toxicity assessments and deepens the understanding of mechanisms driving metal toxicity. Additionally, metabolomics plays a vital role in discovering potential therapeutic compounds and dietary interventions to counteract toxic effects. Recent studies underscore its promise in shaping innovative approaches for managing metal toxicity in both environmental and clinical settings (Bhowmik *et al.*, 2025).

5.4 Nanotechnology in Heavy Metal Remediation

The other promising new front in this area is the application of nanotechnology involving microbial systems to enhance the recovery of HMs and to aid

biomass separation following remediation. Microbially synthesized nanoparticles are emerging as effective agents for remediation of HM-contaminated
waters and soils. All types of microorganisms such as bacteria, fungi and
algae can synthesize nanoparticles by reducing metal ions enzymatically or
via redox metabolites and simultaneously provide capping/stabilizing biomolecules that control nanoparticle size, shape and surface chemistry (Fig. 5.3).
Different biogenic nanoparticles – Ag, gold (Au), Fe_3O_4, FeS, ZnO and CuO
– and mixed-metal/oxide composites have been explored in HM remediation
in different studies and have been found to exhibit high surface areas, active
functional groups and easy magnetic separability. These properties can
enhance the adsorption, reduction or catalytic transformation of toxic metal
species (Altammar, 2023; Jha *et al.*, 2024). To make magnetic nanoparticles
that can be separated easily from treated water, magnetic nanoparticles – particularly those based on iron oxide coated with biocompatible polymers such
as chitosan – have been conjugated to metal-binding bacteria (Gol-Soltani
et al., 2024). To illustrate the high efficiency rate, a synthetic *E. coli* strain
that has been engineered to produce MTs and immobilized onto magnetite
nanoparticles functionalized on chitosan showed superior Cd and Pb removal
capability. This strain reduced concentrations of several milligrams/litre of Cd
and Pb to undetectable limits but with a complete recovery of bacterial cells
by using simple magnetic fields. Not only can such bio-nanohybrids boost
removal efficiency but they can also enable recyclability of the system and help
reduce operational costs (Jafarian and Ghaffari, 2017). Nanocellulose-coated
magnetite modified with *Pseudomonas putida* and *Bacillus megaterium* has been
reported to be more efficient in immobilizing metals in soils than unmodified
nanoparticles, even with low application concentrations, because of enhanced

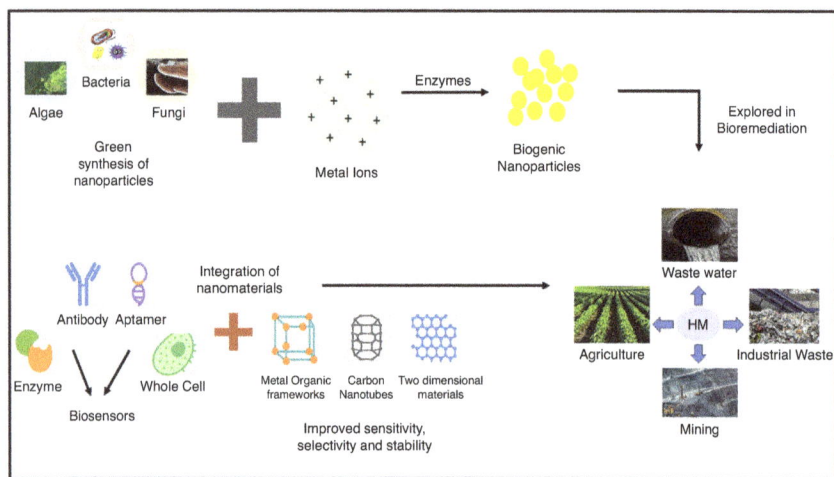

Fig. 5.3. Nanobiotechnology and biosensors for heavy metal bioremediation.

stability and functionalization on the surface (Reshmy *et al.*, 2022). Evidence indicates that engineered nanoparticles such as zero-valent iron and biogenic metal oxides can significantly improve HM immobilization while modulating microbial activity and soil microbiome functions. Recent advances in cultivation, large-scale production and functionalization highlight their vast potential in environmental and industrial applications, making them sustainable and cost-effective tools for pollutant remediation (Yadav *et al.*, 2025).

However, uncertainties remain concerning nanoparticle–microbe–soil interactions, long-term ecotoxicological risks and field-scale feasibility. Future research should emphasize the design of eco-friendly nanoparticles, the development of soil-specific microbial consortia and interdisciplinary approaches that bridge laboratory innovations with practical applications (Kamyab *et al.*, 2025).

5.5   Biosensors in Identifying and Detoxifying Heavy Metals

Simultaneously, synthetic biology technologies have outlined scalable design approaches to design and build sensors and detoxifiers of bacteria capable of sensing and eliminating HMs in polluted media. These biosensors utilize biological elements such as enzymes, antibodies, aptamers and whole cells to specifically recognize and quantify HMs (Fig. 5.3). The integration of nanomaterials, such as carbon nanotubes, metal–organic frameworks and two-dimensional materials, has improved the sensitivity, selectivity and stability of these sensors. For instance, electrochemical aptamer-based biosensors have demonstrated rapid and real-time detection capabilities for various HMs, offering advantages in environmental monitoring applications (Wu *et al.*, 2023).

To tightly regulate biofilm formation, bacteria have evolved sophisticated regulatory networks that maintain cyclic di-guanosine monophosphate (c-di-GMP) homeostasis. This key bacterial messenger plays a role in controlling processes like cell cycling, biofilm formation, adhesion and virulence. To monitor its levels, a genetically encoded biosensor cdiGEBS was developed, based on the c-di-GMP-responsive transcription factor MrkH, which exhibits a 23-fold fluorescence change and detects both low and high c-di-GMP concentrations. Through biofilm formation, bacteria may be able to defend against various external environmental stress conditions (Li *et al.*, 2025).

Biosensors can be applied as live therapy of HM-contaminated sites in the environment. Bacterial strains can now be designed to sense a variety of HMs at once with engineered gene circuits connecting the metals to remediation capabilities. Biosensor strains of *Bacillus subtilis* have been developed to detect and remove up to 99% of toxic HMs such as Pb, Hg and Cu in 12 hour of exposure. Increased specificity has also been facilitated by advances in transcription factor engineering. Variants of the Pb-responsive regulator PbrR now select Pb in preference to competing ions (e.g. Zn, Cd) at a particular nanomolar concentration when paired with a fluorescent or colorimetric reporter (Hynninen *et al.*, 2010; Hui *et al.*, 2022). In addition, As biosensors have been developed utilizing idealized ars operon promoters with a detection limit of 1.8 nM, which is comparative with detectors with a detection limit of 0.1 mM. Other biosensor platforms are incorporated within micro-fluidic micro-systems or micro-biosensor fuel cell systems, and offer the potential to provide portable solutions that could be reused to offer long-term monitoring of HM-contaminated locations. Taken collectively, these developments underline the potential of genetic engineering, synthetic design and nanotechnology to develop next-generation bacterial platforms that can periodically, sustainably and intelligently decontaminate HMs in various environmental matrices (Chen and Rosen, 2014; Kylilis *et al.*, 2019).

5.6 Microbial Consortia and Biofilm Engineering in Heavy Metal Remediation

Recently, research on bioremediation has turned more towards synergies, in which a combination of different biological systems are employed to overcome the constraints inherent to single-organism models and to reach larger efficiencies in eliminating metals across a wide range of environmental changes. Microbial consortia, comprising diverse microorganisms, have emerged as promising agents for bioremediation due to their synergistic interactions and enhanced metabolic capabilities. Synergistic mechanisms of biofilm-based methods provide potential for improved metal removal. Studies have shown that consortia of metal-tolerant bacteria demonstrated greater tolerance and remediation of metals than single strains of the same bacteria. Qurbani *et al.* (2025) found that the consortium of *Pseudomonas putida* and *Pasteurella aerogenes* exhibited more Cu, Zn and Ni removal rates with greater tolerance than any strain alone. In another study, Khidr *et al.* (2025) reported that a co-culture of *Aeromonas* sp. and *Shewanella* sp. enhanced the reduction of Pb and Cd through metabolic synergy. Qattan (2025) demonstrated that mixed indigenous bacterial strains were able to remediate Cd, Ni and Cr using sequential mechanisms such as bioaccumulation, biosorption and biotransformation. Wang *et al.* (2024b) demonstrated that extracellular proteins produced by *Cupriavidus pauculus* enhanced Ni resistance and removal efficiency, while engineered microbial biofilms were found by Li *et al.* (2021) to protect consortia and improve HM removal.

A consortium of microorganisms and plants was explored by Zhang *et al.* (2025a) to mitigate plant stress and found decreased availability of HMs and micro-plastics when they were exposed to a combination of synergistic plants and microbes. Microbial–plant synergy takes advantage of the interdependence between plant roots and rhizosphere tissue-bound bacteria to promote metal uptake, detoxification and stabilization (Kuppan *et al.*, 2024). Plants also shed root exudates that are abundant in organic acids, sugars and phenolics that encourage bacterial growth and activity. Bacteria in turn produce siderophores, organic acids and phytohormones that enhance the solubility and mobilization of the metals, growth of the roots and help the bacteria reduce metal toxicity by chelating and transforming them (Chen and Liu, 2024). *Bacillus* sp. and *Cupriavidus* sp. from Cd-contaminated soils significantly enhanced seed germination and seedling growth of *Vigna radiata* and *Cicer arietinum*, roughly doubling performance compared with controls, demonstrating the effectiveness of bacterial isolates for bioremediation in highly Cd-polluted environments (Lata *et al.*, 2021).

Rhizosphere-associated consortia combined with legumes showed synergistic accumulation of Pb, Cd and Zn, as reported by Zheng *et al.* (2023). Mishra *et al.* (2021) demonstrated that *Pseudomonas* sp. associated with sunflower plants can enhance HM uptake in contaminated soils, through phytoremediation. The mixed microbial–soil consortia system is crucial for the effective degradation of metal pollutants such as Pb, Cr and Ni, according to Zhang *et al.* (2022), with the microbial partner playing an important role in phytoremediation of the HMs. Sanjana *et al.* (2024) also reported that when applied in mixed microbial consortia, a significant improvement in phytoremediation efficiency was reported for HMs such as Cd, Zn and Pb. Microbial consortia composed of *Pseudomonas* sp., *Bacillus* sp. and *Rhizobium* sp. can solubilize HMs such as Cd and Pb to improve uptake by hyperaccumulator plants like *Brassica juncea* and *Helianthus annuus*, increasing phytoremediation rates. Another promising mixed strategy is using consortia of bacteria and fungi, particularly in acidic and extreme environments where single strains of microbes cannot achieve this level of bioremediation (Qin *et al.*, 2024). Powerful organic acids produced by some fungi such as *Aspergillus* sp. and *Penicillium* sp. dissolve metal precipitates and release cations into solution, and the metals are removed by bacteria in the consortium by adsorption of metal or bioaccumulation of metal or they can be biotransformed through the action of enzymes. Not only can this complementary relationship increase rates of bioremediation of HMs from contaminated sites but also improve the stability of microbes to variation in environmental factors such as pH and temperature (Karaffa *et al.*, 2021; Ibrahim *et al.*, 2025).

These studies demonstrate that microbial consortia through their synergistic interactions contribute to improved tolerance, survival and remediation efficiency for HMs. The further use of efficacious microbial agents in soil-amendment techniques, when used together, enhanced end products, which makes microbial consortia very valuable tools for bioremediation of HM-contaminated sites.

Another combined strategy uses bacterial biofilms in conjunction with the biomineralization processes to fix the metals in permanently insoluble forms (Fig. 5.4). Biofilms – complex communities of microorganisms anchored within EPS – provide high-density binding sites to cations through their anionic functional groups (carboxyl, phosphate and sulfate). When coupled with biomineralization routes, including microbially induced carbonate precipitation or sulfate reduction, such systems transform water-soluble metal ions into very stable mineral forms such as metal carbonates, phosphates

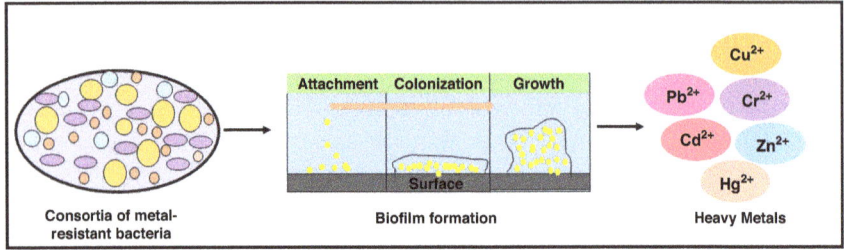

Fig. 5.4. Biofilm-mediated heavy metal remediation by bacterial consortia.

or sulfides (Zhao *et al.*, 2023). As an example, hydrogen sulfide produced by sulfate-reducing bacteria (*Desulfovibrio* sp., *Desulfobulbus* sp.) is used to precipitate Cd and Pb to CdS and PbS precipitates. The generation of carbonate is used to immobilize Cr and Zn and is readily co-precipitated with $CaCO_3$ in soil matrices by *Bacillus* sp. In addition to minimizing the bioavailability of metals, these combined strategies result in minimal secondary contamination and further stabilize remediation sites, which are crucial elements of novel, environmentally friendly bioremediation solutions (Xu and Chen, 2020).

5.7   Limitations and Challenges

Although advances have been made with bioremediation technologies, there are some limitations and challenges that affect the efficiency of bioremediation and subsequent long-term treatment strategies. Increased levels of HMs inhibit the performance of microbial systems as they make the environment prohibitive to microbes by affecting their metabolism, they cause oxidative stress and inhibit bioremediation. Even metal-tolerant strains of microbes will not remain viable in the presence of multi-metal contaminants. Metal toxicity can be synergistic, which has serious implications for studying the impacts of any anthropogenic contamination (Mehrotra *et al.*, 2021). Potential horizontal gene transfer of metal resistance genes in a mixed microbial ecology, such as soil or sewage, is a significant concern as it unintentionally allows for the transfer of resistance traits. For example, if the metal-resistant strain proliferates throughout the ecosystem, it can cause pathogenic effects on human health and ecology. The predominant limiting factor for scaling up bioremediation systems from laboratory to field scales is the lack of engineering principles and operational controls. If left alone, the environmental context may create challenges that will far exceed the possibility of creating or maintaining operational control over factors such as optimal pH, temperature, aeration and nutrient flow. In large-scale processes, energy inputs are high and there may be excessive operational controls and unprecedented demands on bioreactor systems (Aminov, 2011). Furthermore, contextual and experiential factors related to post-treatment biomass recovery, secondary contamination and legislative compliance can create many barriers to increasing bioremediation.

All of these factors suggest the need to develop new, mixed approaches to more explicitly bring together extractive microbial biotechnology with process modelling to enable the development of delivery, immobilization matrices and genetic engineering into workable, scalable and safe remedial solutions (Zhang *et al.*, 2025a).

Bioremediation through the augmentation of genetically engineered microorganisms presents ecological risks. The introduction of engineered organisms into the ecosystem is one of the most carefully registered interventions with a regulatory approval process designed to meet ethical obligations from multiple entities. The risks from engineered microbial traits, interacting with native microbial communities and other life forms, could cause instability of the existing ecological network. Additionally, engineered microbes that are designed for specific niches, such as soil or freshwater, may encounter native populations in a way that could disrupt biodiversity and functional ecosystem processes in environments that may not be fully understand or are difficult to predict (CBAN, 2023; Friends of the Earth, 2023; George and Wan, 2023).

The incorporation of nanomaterials into wastewater treatment systems has also raised concerns over toxicity in aquatic systems due to their size and large surface area. These properties of nanoparticles can magnify their effect on other organisms and can disturb the latter's normal activities. It has been reported that nanomaterials can generate ROS that can induce oxidative stress in aquatic biological organisms that affect cellular activity and can generate apoptosis in tissues. For example, Ag and Cu oxide nanoparticles exhibit anti-bacterial properties, however when released in water, Ag and Cu are capable of destroying microbial communities that are involved in promoting the health of aquatic systems. There are concerns around environmental persistence and transformation of the nanomaterials, which could lead to their accumulation in environments and possible biomagnifications, creating challenging future risks (Jackson *et al.*, 2025). Another obstacle to bioremediation is applying a laboratory research developmental process to the field. Guidelines are needed to be able to replicate the scaling-up process, as well, to be able to ensure results can be replicated when going to uncontrolled field conditions (Erakca *et al.*, 2024).

5.8 Conclusion and Future Prospects

The bioremediation of HMs has the potential opportunity to be developed further with regards to performance, flexible structure and sustainability through genetic and computational techniques in biotechnology. A proposed and tested solution is through a tailored microbial consortium to remediate multi-metals. Consistently pooling genetically optimized bacteria with complementary metabolic pathways, these bacterial consortia can work synergistically together to remediate difficult mixtures of metals at contaminated sites (Rathour *et al.*, 2024). These types of consortia can potentially hold high biosorption capacity along with the ability to perform enzymatic transformation and can be resiliently capable of biofilm formation. This expands persistence in the consortia in a more dynamic environment. A second evolution in science is the application of artificial intelligence (AI) and predictive model tools to better understand bacterial behaviour and the kinetics of metal binding, as well as the dynamics of systems in nature (Mondal *et al.*, 2025).

New ways to utilize AI and machine learning could be useful in optimizing the arrangement of bioreactors, to predict the effectiveness of bioremediation in challenging conditions (e.g. changing the bioreactor's pH or temperature modes), to facilitate genetic controls and more accurately aim for remediation by removing experimentation and testing altogether. Similarly, as bioremediation methods are developed, a circular economy consideration for metal recovery and the reuse of biomass following biosorption is emerging (Blessing and Olateru, 2025). To transform waste biomass into a resource capable of industrial reuse, advanced desorption methods and bio-electrochemical systems are being developed to recover valuable metals, including Cu, Co and rare earth components, out of fed bacterial biomass. Not only do those solutions prove beneficial as a bioremediation approach because of being more economically feasible but they can also contribute to the sustainability of the resources in the environment as well. All these inventions tend to be associated with next-generation innovative, skilled and viable bioremediation technologies which are co-ordinated towards higher aspirations of sustainability (Nancharaiah *et al.*, 2015).

References

Algahmadi, A., Mohammed, A.E., Alfadda, A.A., Alanazi, I.O., Alwehaibi, M.A. *et al.* (2024) Proteomics of *Penicillium chrysogenum* for a deeper understanding of lead (Pb) metal bioremediation. *ACS Omega* 9(24), 26245–26256. DOI: 10.1021/acsomega.4c02006.

Alka, S., Shahir, S., Ibrahim, N., Rahmad, N., Haliba, N. *et al.* (2021) Histological and proteome analyses of *Microbacterium foliorum*-mediated decrease in arsenic toxicity in *Melastoma malabathricum*. *3 Biotech* 11(7), 336. DOI: 10.1007/s13205-021-02864-y.

Altammar, K.A. (2023) A review on nanoparticles: Characteristics, synthesis, applications, and challenges. *Frontiers in Microbiology* 14, 1155622. DOI: 10.3389/fmicb.2023.1155622.

Aminov, R.I. (2011) Horizontal gene exchange in environmental microbiota. *Frontiers in Microbiology* 2, 158. DOI: 10.3389/fmicb.2011.00158.

Barkay, T., Miller, S.M. and Summers, A.O. (2003) Bacterial mercury resistance from atoms to ecosystems. *FEMS Microbiology Reviews* 27(2–3), 355–384.

Ben Fekih, I., Zhang, C., Li, Y.P., Zhao, Y., Alwathnani, H.A. *et al.* (2018) Distribution of arsenic resistance genes in prokaryotes. *Frontiers in Microbiology* 9, 2473.

Bhowmik, S., Kumar, S., Raj, A., Adom, D. and Saxena, R. (2025) Metabolomics in understanding and mitigating metal toxicity. In: *Microbial Metabolomics: Recent*

Developments, Challenges and Future Opportunities. Springer Nature Singapore, Singapore, pp. 347–375.

Blessing, A.A. and Olateru, K. (2025) AI-driven optimization of bioremediation strategies for river pollution: A comprehensive review and future directions. *Frontiers in Microbiology* 16, 1504254. DOI: 10.3389/fmicb.2025.1504254.

Bondarczuk, K. and Piotrowska-Seget, Z. (2013) Molecular basis of active copper resistance mechanisms in gram-negative bacteria. *Cell Biology and Toxicology* 29(6), 397–405. DOI: 10.1007/s10565-013-9262-1.

Borremans, B., Hobman, J.L., Provoost, A., Brown, N.L. and Lelie, D. (2001) Cloning and functional analysis of the pbr lead resistance determinant of *Ralstonia metallidurans* CH34. *Journal of Bacteriology* 183(19), 5651–5658.

Branco, R., Chung, A.P. and Morais, P.V. (2008) Sequencing and expression of two arsenic resistance operons with different functions in the highly arsenic-resistant strain *Ochrobactrum tritici* SCII24T. *BMC Microbiology* 8(1), 95.

Burbano, D.A., Kiattisewee, C., Karanjia, A.V., Cardiff, R.A.L., Faulkner, I.D. *et al.* (2024) CRISPR tools for engineering prokaryotic systems: Recent advances and new applications. *Annual Review of Chemical and Biomolecular Engineering* 15(1), 389–430. DOI: 10.1146/annurev-chembioeng-100522-114706.

Capdevila, D.A., Rondón, J.J., Edmonds, K.A., Rocchio, J.S., Dujovne, M.V. *et al.* (2024) Bacterial metallostasis: Metal sensing, metalloproteome remodeling, and metal trafficking. *Chemical Reviews* 124(24), 13574–13659. DOI: 10.1021/acs.chemrev.4c00264.

CBAN (Canadian Biotechnology Action Network) (2023) *Environmental Impacts of Genetically Modified Organisms*. Ottawa: CBAN.

Chen, J. and Rosen, B.P. (2014) Biosensors for inorganic and organic arsenicals. *Biosensors* 4(4), 494–512. DOI: 10.3390/bios4040494.

Chen, L. and Liu, Y. (2024) The function of root exudates in the root colonization by beneficial soil rhizobacteria. *Biology* 13(2), 95. DOI: 10.3390/biology13020095.

Chen, Y., Cheng, M., Feng, X., Niu, X., Song, H. *et al.* (2022) Genome editing by CRISPR/ Cas12 recognizing AT-Rich PAMs in *Shewanella oneidensis* MR-1. *ACS Synthetic Biology* 11(9), 2947–2955. DOI: 10.1021/acssynbio.2c00208.

Chourey, K., Thompson, M.R., Morrell-Falvey, J., Verberkmoes, N.C., Brown, S.D. *et al.* (2006) Global molecular and morphological effects of 24-hour chromium(VI) exposure on *Shewanella oneidensis* MR-1. *Applied and Environmental Microbiology* 72(9), 6331–6344. DOI: 10.1128/AEM.00813-06.

D'Ermo, G., Audebert, S., Camoin, L., Planer-Friedrich, B., Casiot-Marouani, C. *et al.* (2024) Quantitative proteomics reveals the sox system's role in sulphur and arsenic metabolism of phototroph *Halorhodospira halophila*. *Environmental Microbiology* 26(6), e16655. DOI: 10.1111/1462-2920.16655.

De Silva, M., Cao, G. and Tam, M.K. (2025) Nanomaterials for the removal and detection of heavy metals: A review. *Environmental Science* 16, e16655. DOI: 10.1111/1462-2920.16655.

Erakca, M., Baumann, M., Helbig, C. and Weil, M. (2024) Systematic review of scale-up methods for prospective life cycle assessment of emerging technologies. *Journal of Cleaner Production* 451, 142161. DOI: 10.1016/j.jclepro.2024.142161.

Eshawu, A.B. and Ghalsasi, V.V. (2024) Metabolomics of natural samples: A tutorial review on the latest technologies. *Journal of Separation Science* 47(1), e2300588. DOI: 10.1002/jssc.202300588.

Fang, Y., Yang, Q., Mu, K., Wang, Q., Liu, K. *et al.* (2024) A promotion strategy of enhancing the mercury removal in *Shewanella oneidensis* MR-1 based on the mercury absorption and electronic consumption via mer operon. *Journal of Environmental Chemical Engineering* 12(3), 112993. DOI: 10.1016/j.jece.2024.112993.

Feng, L., Liu, X., Wang, N., Shi, Z., Wang, Y. *et al.* (2025) Genomic analysis of cadmium-resistant and plant growth-promoting *Burkholderia alba* isolated from plant rhizosphere. *Agronomy* 15(8), 1780. DOI: 10.3390/agronomy15081780.

Friends of the Earth (2023) Genetically Engineered Soil Microbes: Risks and Concerns. *Friends of the Earth*, Washington, DC,

Galisteo, C., Puente-Sánchez, F., Haba, R.R., Bertilsson, S., Sánchez-Porro, C. *et al.* (2024) Metagenomic insights into the prokaryotic communities of heavy metal-contaminated hypersaline soils. *Science of the Total Environment* 951, 175497. DOI: 10.1016/j.scitotenv.2024.175497.

George, S.E. and Wan, Y. (2023) Microbial functionalities and immobilization of environmental lead: Biogeochemical and molecular mechanisms and implications for bioremediation. *Journal of Hazardous Materials* 457, 131738.

Gillieatt, B.F. and Coleman, N.V. (2024) Unravelling the mechanisms of antibiotic and heavy metal resistance co-selection in environmental bacteria. *FEMS Microbiology Reviews* 48(4), fuae017. DOI: 10.1093/femsre/fuae017.

Gol-Soltani, M., Ghasemi-Fasaei, R., Ronaghi, A., Zarei, M., Zeinali, S. *et al.* (2024) Efficient immobilization of heavy metals using newly synthesized magnetic nanoparticles and some bacteria in a multi-metal contaminated soil. *Environmental Science and Pollution Research International* 31(27), 39602–39624. DOI: 10.1007/s11356-024-33808-7.

Gupta, A., Matsui, K., Lo, J.F. and Silver, S. (1999) Molecular basis for resistance to silver cations in *Salmonella*. *Nature Medicine* 5(2), 183–188.

Gupta, N., Yadav, K.K., Kumar, V., Cabral-Pinto, M.M., Alam, M. *et al.* (2021) Appraisal of contamination of heavy metals and health risk in agricultural soil of Jhansi city, India. *Environmental Toxicology and Pharmacology* 88, 103740.

Hasman, H. and Aarestrup, F.M. (2002) TcrB, a gene conferring transferable copper resistance in *Enterococcus faecium*: Occurrence, transferability, and linkage to macrolide and glycopeptide resistance. *Antimicrobial Agents and Chemotherapy* 46(5), 1410–1416.

Hassan, M.E.T., Lelie, D., Springael, D., Romling, U., Ahmed, N. *et al.* (1999) Identification of a gene cluster, czr, involved in cadmium and zinc resistance in *Pseudomonas aeruginosa*. *Gene* 238, 417–425.

He, N., Yao, W. and Tang, L. (2024) Surface expression of metallothionein enhances bioremediation in *Escherichia coli*. *Desalination and Water Treatment* 317, 100070. DOI: 10.1016/j.dwt.2024.100070.

He, N., Wang, Z., Lei, L., Chen, C., Qin, Y. *et al.* (2025) Enhancing high-efficient cadmium biosorption of *Escherichia coli* via cell surface displaying metallothionien CUP1. *Environmental Technology* 46(7), 1021–1030. DOI: 10.1080/09593330.2024.2375006.

He, Y., Wang, L., Ma, W., Lu, X., Li, Y. *et al.* (2019) Secretory expression, immunoaffinity purification and metal-binding ability of recombinant metallothionein (ShMT) from freshwater crab *Sinopotamon henanense*. *Ecotoxicology and Environmental Safety* 169, 457–463. DOI: 10.1016/j.ecoenv.2018.11.065.

Hui, C.Y., Guo, Y., Li, H., Chen, Y.T. and Yi, J. (2022) Differential detection of bioavailable mercury and cadmium based on a robust dual-sensing bacterial biosensor. *Frontiers in Microbiology* 13, 846524. DOI: 10.3389/fmicb.2022.846524.

Hynninen, A., Tönismann, K. and Virta, M. (2010) Improving the sensitivity of bacterial bioreporters for heavy metals. *Bioengineered Bugs* 1(2), 132–138. DOI: 10.4161/bbug.1.2.10902.

Ibrahim, A., Oginga, B., Zhang, Y., Ling, W., Tang, L. *et al.* (2025) Bioremediation of soils with emerging organic contaminants using immobilized microorganisms. *Environmental Technology and Innovation* 40, 104345. DOI: 10.1016/j.eti.2025.104345.

Jackson, J.S., Kantamaneni, K., Ganeshu, P., Sunkur, R. and Rathnayake, U. (2025) Assessment of the role of nanotechnology in water sector: An expert opinion. *International Journal of Energy and Water Resources* 9, 1645–1667. DOI: 10.1007/s42108-025-00389-1.

Jafarian, V. and Ghaffari, F. (2017) A unique metallothionein-engineered in *Escherichia coli* for biosorption of lead, zinc, and cadmium; absorption or adsorption? *Microbiology* 86(1), 73–81. DOI: 10.1134/S0026261717010064.

Jha, A., Barsola, B., Pathania, D., Raizada, P. and Thakur, P. (2024) Nano-biogenic heavy metals adsorptive remediation for enhanced soil health and sustainable agricultural production. *Environmental Research* 252(Pt 3), 118926. DOI: 10.1016/j.envres.2024.118926.

Jiang, Z., Hong, X., Zhang, S., Yao, R. and Xiao, Y.I. (2020) CRISPR base editing and prime editing: DSB and template-free editing systems for bacteria and plants. *Synthetic and Systems Biotechnology* 5(4), 277–292. DOI: 10.1016/j.synbio.2020.08.003.

Kamyab, H., Chelliapan, S., Khalili, E., Mediavilla, D.P.Z., Khorami, M. *et al.* (2025) Nanobioremediation of heavy metals using microorganisms. *Journal of Environmental Management* 392, 126736. DOI: 10.1016/j.jenvman.2025.126736.

Karaffa, L., Fekete, E. and Kubicek, C.P. (2021) The role of metal ions in fungal organic acid accumulation. *Microorganisms* 9(6), 1267. DOI: 10.3390/microorganisms9061267.

Kaur, R., Kaur, S., Singh, R. and Nair, M. (2023) Proteomics in shaping the future of biofertiliser delivery technique. In: *Metabolomics, Proteomes and Gene Editing Approaches in Biofertilizer Industry*. Springer Nature Singapore, Singapore, pp. 325–337.

Kavya, T., Kumari, H.K., Singh, G., Govindasamy, V., Vijaysri, D. *et al.* (2024) *Gene Editing Tools for Engineering Beneficial Microorganism in Biofertilizer*. Springer Nature Singapore, Singapore, pp. 83–98.

Khidr, R., Qurbani, K., Muhammed, V., Salim, S., Abdulla, S. *et al.* (2025) Synergistic effects of indigenous bacterial consortia on heavy metal tolerance and reduction. *Environmental Geochemistry and Health* 47(3), 79.

Kuppan, N., Padman, M., Mahadeva, M., Srinivasan, S. and Devarajan, R. (2024) A comprehensive review of sustainable bioremediation techniques: Eco friendly solutions for waste and pollution management. *Waste Management Bulletin* 2(3), 154–171. DOI: 10.1016/j.wmb.2024.07.005.

Kylilis, N., Riangrungroj, P., Lai, H.E., Salema, V., Fernández, L.Á. *et al.* (2019) Whole-cell biosensor with tunable limit of detection enables low-cost agglutination assays for medical diagnostic applications. *ACS Sensors* 4(2), 370–378. DOI: 10.1021/acssensors.8b01163.

Lata, S., Mishra, T. and Kaur, S. (2021) Cadmium bioremediation potential of *Bacillus* sp. and *Cupriavidus* sp. *Journal of Pure and Applied Microbiology* 15(3), 1665–1680. DOI: 10.22207/JPAM.15.3.63.

Lata, S., Sharma, S. and Kaur, S. (2023) OMICS approaches in mitigating metal toxicity in comparison to conventional techniques used in cadmium bioremediation. *Water, Air, and Soil Pollution* 234(3), 148. DOI: 10.1007/s11270-023-06145-7.

Lehembre, F., Doillon, D., David, E., Perrotto, S., Baude, J. *et al.* (2013) Soil metatranscriptomics for mining eukaryotic heavy metal resistance genes. *Environmental Microbiology* 15(10), 2829–2840.

Li, H., Quan, S. and He, W. (2025) A genetically encoded fluorescent biosensor for sensitive detection of cellular c-di-GMP levels in *Escherichia coli*. *Frontiers in Chemistry* 12, 1528626. DOI: 10.3389/fchem.2024.1528626.

Liu, W., Li, L., Jiang, J., Wu, M. and Lin, P. (2021) Applications and challenges of CRISPR-Cas gene-editing to disease treatment in clinics. *Precision Clinical Medicine* 4(3), 179–191. DOI: 10.1093/pcmedi/pbab014.

Li, X., Wu, S., Dong, Y., Fan, H., Bai, Z. *et al.* (2021) Engineering microbial consortia towards bioremediation. *Water* 13(20), 2928.

Li, Y., Shi, X., Chen, Y., Luo, S., Qin, Z. *et al.* (2023) Quantitative proteomic analysis of the mechanism of Cd toxicity in *Enterobacter* sp. FM-1: Comparison of different growth stages. *Environmental Pollution* 336, 122513. DOI: 10.1016/j.envpol.2023.122513.

Liang, Y., Jiao, S., Wang, M., Yu, H. and Shen, Z. (2020) A CRISPR/Cas9-based genome editing system for *Rhodococcus ruber* TH. *Metabolic Engineering* 57, 13–22. DOI: 10.1016/j.ymben.2019.10.003.

Liu, Y., Yu, W., Nie, T., Wang, L. and Niu, Y. (2025) Extracellular Cr(VI) reduction by the salt-tolerant strain *Bacillus safensis* BSF-4. *Microorganisms* 13(8). DOI: 10.3390/microorganisms13081961.

Lu, C.-W., Ho, H.-C., Yao, C.-L., Tseng, T.-Y., Kao, C.-M. *et al.* (2023) Bioremediation potential of cadmium by recombinant *Escherichia coli* surface expressing metallothionein MTT5 from *Tetrahymena thermophila*. *Chemosphere* 310, 136850. DOI: 10.1016/j.chemosphere.2022.136850.

Marrero, J., Auling, G., Coto, O. and Nies, D.H. (2007) High-level resistance to cobalt and nickel but probably no transenvelope efflux: Metal resistance in the Cuban *Serratia marcescens* strain C-1. *Microbial Ecology* 53(1), 123–133.

Mehrotra, T., Dev, S., Banerjee, A., Chatterjee, A., Singh, R. *et al.* (2021) Use of immobilized bacteria for environmental bioremediation: A review. *Journal of Environmental Chemical Engineering* 9(5), 105920. DOI: 10.1016/j.jece.2021.105920.

Mishra, S., Lin, Z., Pang, S., Zhang, Y., Bhatt, P. *et al.* (2021) Biosurfactant is a powerful tool for the bioremediation of heavy metals from contaminated soils. *Journal of Hazardous Materials* 418, 126253.

Mohammad, S.J., Ling, Y.E., Halim, K.A., Sani, B.S. and Abdullahi, N.I. (2025) Heavy metal pollution and transformation in soil: A comprehensive review of natural bioremediation strategies. *Journal of Umm Al-Qura University for Applied Sciences* 11(3), 528–544. DOI: 10.1007/s43994-025-00241-6.

Mondal, S., Melzi, A., Zecchin, S. and Cavalca, L. (2025) Quorum sensing in biofilm-mediated heavy metal resistance and transformation: Environmental perspectives and bioremediation. *Frontiers in Microbiology* 16, 1607370. DOI: 10.3389/fmicb.2025.1607370.

Mukherjee, A. and Reddy, M.S. (2020) Metatranscriptomics: An approach for retrieving novel eukaryotic genes from polluted and related environments. *3 Biotech* 10, 1–19.

Mukherjee, A., Yadav, R., Marmeisse, R., Fraissinet-Tachet, L. and Reddy, M.S. (2019) Heavy metal hypertolerant eukaryotic aldehyde dehydrogenase isolated from metal contaminated soil by metatranscriptomics approach. *Biochimie* 160, 183–192.

Naiel, M.A.E., Taher, E.S., Rashed, F., Ghazanfar, S., Shehata, A.M. *et al.* (2024) The arsenic bioremediation using genetically engineered microbial strains on aquatic environments: An updated overview. *Heliyon* 10(17), e36314. DOI: 10.1016/j.heliyon.2024.e36314.

Nancharaiah, Y.V., Venkata Mohan, S. and Lens, P.N.L. (2015) Metals removal and recovery in bioelectrochemical systems: A review. *Bioresource Technology* 195, 102–114. DOI: 10.1016/j.biortech.2015.06.058.

Nies, D.H. (1995) The cobalt, zinc, and cadmium efflux system CzcABC from *Alcaligenes eutrophus* functions as a cation-proton antibporter in *Eschericia coli*. *Journal of Bacteriology* 177, 2707–2712.

Pal, C., Bengtsson-Palme, J., Rensing, C., Kristiansson, E. and BacMet, D.L. (2014) Antibacterial biocide and metal resistance genes database. *Nucleic Acids Research* 42, D737–D743. DOI: 10.1093/nar/gkt1252.

Pei, Y., Tao, C., Ling, Z., Yu, Z., Ji, J. *et al.* (2020) Exploring novel Cr (VI) remediation genes for Cr (VI)-contaminated industrial wastewater treatment by comparative metatranscriptomics and metagenomics. *Science of the Total Environment* 742, 140435. DOI: 10.1016/j.scitotenv.2020.140435.

Peris-Díaz, M.D., Orzeł, A., Wu, S., Mosna, K., Barran, P.E. *et al.* (2024) Combining native mass spectrometry and proteomics to differentiate and map the metalloform landscape in metallothioneins. *Journal of Proteome Research* 23(8), 3626–3637. DOI: 10.1021/acs.jproteome.4c00271.

Qattan, S.Y.A. (2025) Harnessing bacterial consortia for effective bioremediation: Targeted removal of heavy metals, hydrocarbons, and persistent pollutants. *Environmental Sciences Europe* 37(1), 85. DOI: 10.1186/s12302-025-01103-y.

Qin, H., Wang, Z., Sha, W., Song, S., Qin, F. *et al.* (2024) Role of plant-growth-promoting rhizobacteria in plant machinery for soil heavy metal detoxification. *Microorganisms* 12(4), 700. DOI: 10.3390/microorganisms12040700.

Qurbani, K., Wsw, H., Khdhr, R., Hussein, S., Ibrahim, B. *et al.* (2025) Synergistic enhancement of heavy metal tolerance and reduction by indigenous bacterial consortia of *Pseudomonas putida* and *Pasteurella aerogenes*. *Scientific Reports* 15(1), 24663. DOI: 10.1038/s41598-025-99238-8.

Rani, N. and Sagar, N.A. (2024) Metabolomics: A paradigm shift in understanding biofertilizers dynamics. In: *Metabolomics, Proteomics and Gene Editing Approaches in Biofertilizer Industry: Volume II*. Springer Nature Singapore, Singapore, pp. 35–51.

Rashid, M. and Stingl, U. (2015) Contemporary molecular tools in microbial ecology and their application to advancing biotechnology. *Biotechnology Advances* 33(8), 1755–1773. DOI: 10.1016/j.biotechadv.2015.09.005.

Rastkhah, E., Fatemi, F. and Maghami, P. (2024) Optimizing the metal bioreduction process in recombinant *Shewanella azerbaijanica* bacteria: A novel approach via mtr C gene cloning and nitrate-reducing pathway destruction. *Molecular Biotechnology* 66(11), 3150–3163. DOI: 10.1007/s12033-023-00920-x.

Rathour, R.K., Sharma, D., Ullah, S., Mahmoud, E.H.M., Sharma, N. *et al.* (2024) Bacterial–microalgal consortia for bioremediation of textile industry wastewater and resource recovery for circular economy. *Biotechnology for the Environment* 1(1), 6. DOI: 10.1186/s44314-024-00005-2.

Rensing, C. and Grass, G. (2003) *Escherichia coli* mechanisms of copper homeostasis in a changing environment. *FEMS Microbiology Reviews* 27(2–3), 197–213.

Reshmy, R., Philip, E., Madhavan, A., Pugazhendhi, A., Sindhu, R. *et al.* (2022) Nanocellulose as green material for remediation of hazardous heavy metal contaminants. *Journal of Hazardous Materials* 424(Pt B), 127516. DOI: 10.1016/j.jhazmat.2021.127516.

Sanjana, M., Prajna, R., Urvi, S.K. and Kavitha, R.V. (2024) Bioremediation – the recent drift towards a sustainable environment. *Environmental Science: Advances* 3(8), 1097–1110.

Sarker, A., Masud, M.A.A., Deepo, D.M., Das, K., Nandi, R. *et al.* (2023) Biological and green remediation of heavy metal contaminated water and soils: A state-of-the-art review. *Chemosphere* 332, 138861. DOI: 10.1016/j.chemosphere.2023.138861.

Schmidt, T. and Schlegel, H.G. (eds) (1994) Combined nickel-cobalt-cadmium resistance encoded by the ncc locus of *Alcaligenes xylosoxidans* 31A. *Journal of Bacteriology* 176(22), 7045–7054.

Simon, C. and Daniel, R. (2011) Metagenomic analyses: Past and future trends. *Applied and Environmental Microbiology* 77(4), 1153–1161.

Staehlin, B.M., Gibbons, J.G., Rokas, A., O'Halloran, T.V. and Slot, J.C. (2016) Evolution of a heavy metal homeostasis/resistance island reflects increasing copper stress in enterobacteria. *Genome Biology and Evolution* 8(3), 811–826.

Sun, N. and Zhao, H. (2013) Transcription activator-like effector nucleases (TALENs): A highly efficient and versatile tool for genome editing. *Biotechnology and Bioengineering* 110(7), 1811–1821. DOI: 10.1002/bit.24890.

Tang, H., Xiang, G., Xiao, W., Yang, Z. and Zhao, B. (2024) Microbial mediated remediation of heavy metals toxicity: Mechanisms and future prospects. *Frontiers in Plant Science* 15, 1420408. DOI: 10.3389/fpls.2024.1420408.

Thai, T.D., Lim, W. and Na, D. (2023) Synthetic bacteria for the detection and bioremediation of heavy metals. *Frontiers in Bioengineering and Biotechnology* 11, 1178680. DOI: 10.3389/fbioe.2023.1178680.

Thakur, B., Yadav, R., Fraissinet-Tachet, L., Marmeisse, R. and Reddy, M.S. (2018) Isolation of multi-metal tolerant ubiquitin fusion protein from metal polluted soil by metatranscriptomic approach. *Journal of Microbiological Methods* 152, 119–125.

Thakur, B., Kaur, S., Mouafo, H.T., Dhiman, S., Kharb, K. *et al.* (2025) Microbial metabolomics: A futuristic approach in biotechnology. In: *Microbial Metabolomics: Recent Developments, Challenges and Future Opportunities*. Springer Nature, Singapore, pp. 1–29.

Tiwary, K. (2025) Synthetic biology and genetic engineering strategies for microbial and algal bioremediation of heavy metals: A scoping review. *Research Square Preprint*. DOI: 10.21203/rs.3.rs-7175786/v1.

Turco, F., Garavaglia, M., Houdt, R., Hill, P., Rawson, F.J. *et al.* (2022) Synthetic biology toolbox, including a single-plasmid CRISPR-Cas9 system to biologically engineer the electrogenic, metal-resistant bacterium *Cupriavidus metallidurans* CH34. *ACS Synthetic Biology* 11(11), 3617–3628. DOI: 10.1021/acssynbio.2c00130.

Wang, L., Wang, Y., Dai, S. and Wang, B. (2024a) Surface display of multiple metal-binding domains in *Deinococcus radiodurans* alleviates cadmium and lead toxicity in rice. *International Journal of Molecular Sciences* 25(23), 12570. DOI: 10.3390/ijms252312570.

Wang, M., Vollstedt, C., Siebels, B., Yu, H., Wu, X. *et al.* (2024b) Extracellular proteins enhance *Cupriavidus pauculus* nickel tolerance and cell aggregate formation. *Bioresource Technology* 393, 130133.

Wang, S., Sun, Y., Wang, S., Fan, C., Wang, D. *et al.* (2024c) Enhanced biosorption of cadmium ions on immobilized surface-engineered yeast using cadmium-binding peptides. *Frontiers in Microbiology* 15, 1496843. DOI: 10.3389/fmicb.2024.1496843.

Wang, Y., Hu, Y., Liu, Y., Chen, Q., Xu, J. *et al.* (2024d) Heavy metal induced shifts in microbial community composition and interactions with dissolved organic matter in coastal sediments. *Science of the Total Environment* 927, 172003. DOI: 10.1016/j.scitotenv.2024.172003.

White, R.J., Collins, J.E., Sealy, I.M., Wali, N., Dooley, C.M. *et al.* (2017) A high-resolution mRNA expression time course of embryonic development in zebrafish. *Elife* 6, e30860.

Wu, B., Ga, L., Wang, Y. and Ai, J. (2023) Recent advances in the application of bionanosensors for the analysis of heavy metals in aquatic environments. *Molecules* 29(1), 34.

Xu, Y.N. and Chen, Y. (2020) Advances in heavy metal removal by sulfate-reducing bacteria. *Water Science and Technology* 81(9), 1797–1827. DOI: 10.2166/wst.2020.227.

Xue, Y., Li, Y., Li, X., Zheng, J., Hua, D. *et al.* (2024) Arsenic bioremediation in mining wastewater by controllable genetically modified bacteria with biochar. *Environmental Technology and Innovation* 33, 103514. DOI: 10.1016/j.eti.2023.103514.

Yadav, V.K., Choudhary, N., Gnanamoorthy, G., Kumar, P., Gupta, R. *et al.* (2025) Recent advances in the bioremediation of wastewater pollutants by using bacterial magnetic nanoparticles and magnetotactic bacteria. *World Journal of Microbiology and Biotechnology* 41(8), 284. DOI: 10.1007/s11274-025-04447-y.

Yu, Z., Pei, Y., Zhao, S., Kakade, A., Khan, A. *et al.* (2021) Metatranscriptomic analysis reveals active microbes and genes responded to short-term Cr(VI) stress. *Ecotoxicology (London, England)* 30(8), 1527–1537. DOI: 10.1007/s10646-020-02290-5.

Zha, S., Yu, A., Wang, Z., Shi, Q., Cheng, X. *et al.* (2024) Microbial strategies for effective hexavalent chromium removal: A comprehensive review. *Chemical Engineering Journal* 489, 151457.

Zhang, C., Singh, R.P., Yadav, P., Kumar, I., Kaushik, A. *et al.* (2025a) Recent advances in biotechnology and bioengineering for efficient microalgal biofuel production. *Fuel Processing Technology* 270, 108199. DOI: 10.1016/j.fuproc.2025.108199.

Zhang, Y., Zhao, S., Liu, S., Peng, J., Zhang, H. *et al.* (2022) Enhancing the phytoremediation of heavy metals by combining hyperaccumulator and heavy metal-resistant plant growth-promoting bacteria. *Frontiers in Plant Science* 13, 912350.

Zhang, Y., Peng, J., Wang, Z., Zhou, F., Yu, J. *et al.* (2025b) Metagenomic analysis revealed the bioremediation mechanism of lead and cadmium contamination by modified biochar synergized with *Bacillus cereus* PSB-2 in phosphate mining wasteland. *Frontiers in Microbiology* 16, 1529784. DOI: 10.3389/fmicb.2025.1529784.

Zhao, A., Sun, J. and Liu, Y. (2023) Understanding bacterial biofilms: From definition to treatment strategies. *Frontiers in Cellular and Infection Microbiology* 13, 1137947. DOI: 10.3389/fcimb.2023.1137947.

Zheng, K., Liu, Z., Liu, C., Liu, J. and Zhuang, J. (2023) Enhancing remediation potential of heavy metal contaminated soils through synergistic application of microbial inoculants and legumes. *Frontiers in Microbiology* 14, 1272591.

www.ingramcontent.com/pod-product-compliance
Lightning Source LLC
Chambersburg PA
CBHW042315210326

41599CB00038B/7134